T0139775

# Solving Large-Scale Production Scheduling and Planning in the Process Industries

Georgios M. Kopanos · Luis Puigjaner

# Solving Large-Scale Production Scheduling and Planning in the Process Industries

 Springer

Georgios M. Kopanos
Flexciton Ltd.
London, UK

Luis Puigjaner
ETSEIB
Universitat Politècnica de Catalunya
Barcelona, Spain

ISBN 978-3-030-13164-7          ISBN 978-3-030-01183-3    (eBook)
https://doi.org/10.1007/978-3-030-01183-3

This Springer imprint is published by the registered company Springer Nature Switzerland AG
The registered company address is: Gewerbestrasse 11, 6330 Cham, Switzerland

*Life is short, art is vast, the opportunity is instantaneous, the experiment is dangerous, the judgment difficult and uncertain*
Based on *Aphorisms*, Hippocrates
(460–477 B.C.)

*This book is dedicated to my daughter, Elisavet*

Georgios M. Kopanos

# Foreword

The need for effective methods and practical tools for production planning and scheduling of manufacturing facilities in the process and related industries has been recognized for at least 50 years. Indeed, among the earliest industrial applications of computers to support manufacturing decisions was the use of linear programming methodology for refinery production planning. Planning models have been extended to include some level of discrete decisions, modeled with 0–1 variables, and a plethora of such models are now routinely solved using commercial Mixed Integer Linear Programming (MILP) solvers. Despite the early start and considerable success with optimization-based planning applications, the development of robust tools for scheduling and their penetration into practice has been much slower. This has not been for lack of effort; certainly, the process systems engineering community has invested considerable effort, produced hundreds of publications and a steady succession of software implementations. The major barrier to solving scheduling problems is the fact that scheduling involves sequencing and assignment decisions, which at root are combinatorial in nature. As a consequence, the representation of these decisions in terms of 0–1 variable rapidly leads to very large dimensionality problems, whose solution using even the best commercial MILP solvers can require impractically long computation times. However, the understanding to which the community has evolved is that this barrier can be mitigated through the development of innovative MILP problem formulations that effectively take advantage of the specific characteristic of particular scheduling problem types and of tailored solution strategies, such as strong problem decompositions, that can exploit the structure of those formulations while using the robust commercial MILP codes to solve the resulting subproblems. Using this strategy, successes with problems of industrial scope are now being reported, setting the stage for this book.

The authors of this book have over a period of years pursued this strategy with considerable success by developing MILP formulations and solution strategies for several important classes of scheduling problems, including single-stage parallel unit plants, continuous and batch, as well types of multistage semicontinuous and batch plants. They have addressed important specific features, such as product sequence-dependent changeover times and costs and the aggregation of products into product families. They have carefully incorporated the characteristics of specific industrial processes, such as those for producing yogurt, ice cream, and beverages as well as production of some types of pharmaceutical active ingredients. In Chaps. 5 and 8 of this monograph, they have compiled and explained these developments in a systematic fashion. They have done a superb job of making this technology accessible not only to those readers with interests in these specific industries but also to readers involved in other manufacturing sectors which feature similar production structures.

A further challenge for this domain of process systems engineering has been the need for effective approaches for integrating manufacturing decisions made at the planning and scheduling levels. Nominally, the role of production planning is, given product requirements and resource constraints, to determine production targets and associated resource allocations for each of the time periods of a multi-time period planning horizon. These targets then set the constraints for the production schedule, in which the necessary detailed sequencing and resource assignment decisions are determined. For this combination of planning and scheduling decisions to be effective, the targets set at the former level have to be realizable when scheduling details are taken into account, yet have to be tight enough so that the period schedules do not leave an unacceptable level of manufacturing resources idle. The authors of this book have made significant advances toward this integration within the context of the manufacturing sectors that they have investigated. For example, they demonstrate how effective use can be made of combinations of discrete- and continuous-time representations within a combined framework. As shown through the challenging case studies that are reported herein, excellent results have been achieved with applications of realistic industrial scope. These developments are very thoroughly reported in Chaps. 4, 6, and 7. An interesting application of this integration concept is also elaborated in Chap. 9, in which a combined approach to the operational and maintenance planning of a manufacturing line is presented. Although the potential conflicts between the needs for maintenance and demands of manufacturing are easily envisioned, the approach that the authors report shows that these can be balanced and real benefits can be derived from careful integration.

To complement the core material of Parts II and through V, in Part I of the book, the authors accommodate readers less familiar with the planning and scheduling domain by providing a compact overview of the relevant structural details of manufacturing operations that are important to planning and scheduling and offering a succinct review of the state of the available literature. Moreover, for those not familiar with the optimization methods currently in use, they include a tutorial on the basic algorithms as well as the modeling system GAMS and MILP solver CPLEX that are used in the case studies reported. It is also commendable that in the appendices they not only provide data and sample results for two of the comprehensive case studies but also pseudocode for the solution strategy implementation. These appendices will prove quite useful to the reader interested in learning from and adapting these case studies to their own applications.

This book reflects the deep knowledge and practical expertise of the authors in the area of planning and scheduling of batch and semicontinuous manufacturing plants. The chapters of the book draw on the portfolio of outstanding publications of the authors in this domain which has been compiled over a decade or more. The case studies, drawn from their own work and with industrial colleagues, are representative of real applications and can serve as test cases for students and researchers in scheduling methodology. The book is admirably suitable for use in postgraduate courses and, of course, for self-study.

This is an important contribution to the literature of MILP-based approaches to planning and scheduling and for this the authors deserve our thanks.

West Lafayette, USA
July 2018

Prof. G. V. Reklaitis
Purdue University

# Preface

Production planning and scheduling constitute a crucial part of the overall supply chain decision level pyramid. Planning and scheduling activities are concerned with the allocation over time of scarce resources between competing activities to meet these requirements in an efficient fashion. More specifically, the planning function aims to optimize the economic performance of the enterprise, as it matches production to demand in the best possible way. The production scheduling component is of vital importance as it is the layer which translates the economic imperatives of the plan into a sequence of actions to be executed on the plant floor, so as to deliver the optimized economic performance predicted by the higher level plan. Overall, recent research is directed toward finding solutions that enable efficient and accurate handling of problems of large size and increasing complexity. However, there remains significant work to be done on both model enhancements and improvements in solution algorithms, if industrially relevant problems are to be tackled routinely, and software based on these are to be used on a regular basis by practitioners in the field. In addition, new academic developments are mostly tested on complex but relatively small- to medium-size problems. Therefore, the implementation of new production and scheduling approaches in real-life industrial case studies constitutes a challenging task.

This book offers an entirely new conception of Planning and detailed production Scheduling in the Process Industries. It provides the answer to the aforementioned existing gaps by providing methods, tools, and techniques for the efficient solution of large-scale production scheduling and planning problems in the process industries. The reader is guided through abundant illustrations (106), tables (29), and a web link for additional consultation that facilitates reading and understanding. The training is complemented with motivating examples and large-scale industrial applications ranging from continuous, to semicontinuous and batch processes.

It is intended as a textbook for academics (Ph.D., M.Sc.), researchers, and industry decision-makers, who are involved in the design, retrofit design, and evaluation activities of alternative scenarios. Teachers can also greatly benefit from this book in the teaching of advanced courses, and industry professionals are provided—by this book—with the know-how to evaluate and improve existing installations or to design a new one.

London, UK                                                    Georgios M. Kopanos
Barcelona, Spain                                                  Luis Puigjaner
July 2018

# Acknowledgements

Undoubtedly, the work presented in this book would not have been possible without the support of family, friends, and colleagues. We would like to express our deepest gratitude to all of you. Specifically, we are truly indebted to Prof. Michael Georgiadis (Aristotle University of Thessaloniki, Greece) for the collaboration in three of the chapters of this book and his invaluable advice in the content of a large part of the book. We are also grateful to Prof. Antonio Espuña (Universitat Politècnica de Catalunya, Spain), and Prof. Gintaras V. Reklaitis (Purdue University, USA) who through fruitful discussions and healthy criticism helped in shaping further the contents of this book. In addition, we would like to express our gratitude to Prof. Christos T. Maravelias (University of Wisconsin-Madison, USA) for the collaboration in Chap. 4, Prof. Carlos A. Méndez (Universidad Nacional del Litoral, Argentina) for the collaboration in Chap. 8, and Mrs. Nur I. Zulkafli and Mrs. Oluwatosin C. Murele (Cranfield University) for the collaboration in Chap. 9 of this book.

# Contents

# Acronyms

| | |
|---|---|
| AI | Artificial Intelligence |
| AM | Agile Manufacturing practices |
| ASP | Approximate Scheduling Formulation |
| B&B | Branch & Bound |
| B+BS | Batching/Batch Scheduling |
| BIL | Bi-Level approach |
| CEFIC | European Chemical Industry Council |
| CP | Constraint Programming paradigm |
| CPLEX | Optimization Solver Package |
| CPU | Central Processing Unit |
| DSP | Detailed Scheduling Formulation |
| EON | Event Operation Network representation |
| EU | European Union |
| FIS | Finite Intermediate Storage policy |
| FP | Final Products |
| GAMS | General Algebraic Modeling System |
| GP | General/Global Precedence approach |
| HMRM | Heuristic Model Reduction Methods |
| I/LP | Immediate/Local Precedence approach |
| IP | Intermediate Products |
| JIT | Just-In-Time management practice |
| KRI-KRI | Dairy Production Facility |
| LM | Lean Management practice |
| LP | Linear Programming |
| MIP | Mixed Integer Programming |
| NIS | No Intermediate Storage policy |
| OCC | Operating Changeover Costs |
| OF | Objective Function |
| OR | Operational Research |
| PACK | Packing line |

| | |
|---|---|
| PLT | Party Logistics Truck |
| PROC | Processing line |
| PSE | Process Systems Engineering |
| RAM | Random Access Memory |
| RM | Raw Materials |
| RTN | Resource Task Network representation |
| SC | Supply Chain |
| SCM | Supply Chain Management |
| S-graph | Representation of a sequence of tasks/recipes |
| SIS | Shared Intermediate Storage policy |
| STN | State Task Network representation |
| TGH | Time Grid Heuristic |
| TOP | Operation time |
| TW | Time Waiting |
| UIS | Unlimited Intermediate Storage policy |
| USGP | Unit-Specific General Precedence approach |
| USIP | Unit-Specific Immediate Precedence approach |
| W-K | Westenberger-Kallrath benchmark scheduling problem |
| WSPT | Weighted Shortest Processing Time |
| ZW | Zero-Wait policy |

# Part I
# Overview

# Chapter 1
# Introduction

## 1.1 Introduction to Manufacturing Types

Nowadays, the two major manufacturing disciplines are process manufacturing and discrete manufacturing. In contrast with discrete manufacturing, process manufacturing results in final (and often intermediate) products that cannot be disassembled back into their original components. For instance, a can of soda cannot be returned to its basic components such as carbonated water, citric acid, potassium benzoate, aspartame, and other ingredients. Orange juice cannot be put back into an orange. However, a car or computer, on the other hand, can be disassembled and its components, to a large extent, returned to stock. In discrete manufacturing, the manufacturing floor works off shop orders to build something and the individual products are easily identifiable. The automobile, the computer, and the aerospace industry are some representative discrete industries. Process manufacturing is common in the food, beverages, chemicals, pharmaceuticals, petroleum, ceramics, and paper industries.

Roughly speaking, machines that do the main processing operations typically have very high startup and shutdown costs and usually work around the clock. A machine in the process industries also incurs a high changeover cost when it has to switch over from one product to another. In process manufacturing, products are undifferentiated. Process manufacturing is the branch of manufacturing that is associated with formulas and manufacturing recipes, and can be contrasted with discrete manufacturing, which is concerned with bills of material and routing. This is more than a subtle difference in terminology; the terms characterize distinct manufacturing approaches. This book is focused on process manufacturing with emphasis on food, pharmaceutical, and chemical products.

© Springer Nature Switzerland AG 2019
G. M. Kopanos and L. Puigjaner, *Solving Large-Scale Production Scheduling and Planning in the Process Industries*,
https://doi.org/10.1007/978-3-030-01183-3_1

## 1.2   Overview of Process Industry and Motivation

According to the European Chemical Industry Council (2017), the European Union (EU) still holds the third largest chemicals producing area in the world. It provides 1,155,000 jobs and helps in generating wealth for us all. However, the spectacular growth rate in China, 12.4% average, during the period 2006–2016 has caused an almost negative growth for Europe, which lost its world second place. However, the perspective is a continued growth, but at slower rate. Europe's major strength remains in the chemical industry's pioneering technologies and efficiency, which wins customers around the world. It is worth noticing that the EU is the world's top exporter and importer of chemicals, accounting for more than 40% of total global trade in 2009. The EU chemical exports in 2016 reached €146.2 billion, while imports from non-EU countries had a value of €99.0 billion during the same period. It is also noteworthy that of the 30 largest chemical companies in the world, 12 are headquartered in Europe (e.g., BASF, Bayer, Shell, Ineos, Total) representing around 10% of world chemical sales. Pharmaceuticals and chemicals form the first- and second-leading EU manufacturing sectors in terms of value-added per employee in 2006, respectively. It is noticeable that food and beverage industries hold the seventh place in this list.

The significant role of the process industry in the EU's economic status is evident. EU's process industry ought to be active and improve its operational and functional performance through its entire supply chain network, in order to maintain its leading position in the world's highly competitive market. The faster growth rhythm of Asian countries, especially China and Japan, has created a strong competitor for EU process industries, thus making indispensable the enhancement of the production management and the overall supply chain network.

### 1.2.1   Supply Chain Management

According to Min and Zhou (2002), Supply Chain (SC) is referred to as an integrated system which synchronizes a series of inter-related business processes in order to (i) acquire raw materials and parts, (ii) transform these raw materials and parts into final products, (iii) add value to these products, (iv) distribute and promote final products to either retailers or customers, and (v) facilitate information exchange among various business entities (i.e., vendors, manufacturers, distributors, third-party logistics providers, and retailers). SC networks consist of a number of echelons (e.g., suppliers, production plants, warehouses, distribution centers, and markets) and are described by a forward flow of materials and a backward flow of information. The SC is often represented as a network similar to the one illustrated in Fig. 1.1.

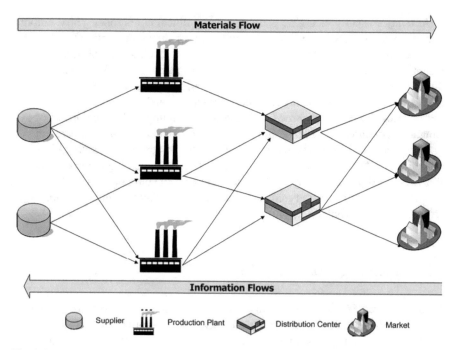

**Fig. 1.1** A typical supply chain network

According to the Council of Supply Chain Management Professionals (2018), Supply Chain Management (SCM) is the process of planning, implementing, and controlling the SC operations in an efficient way. SCM spans all movements (mainly logistics operations) and storage of raw materials, work-in-process inventory, intermediate products, and finished goods from the point-of-origin to the point-of-consumption (Simchi-Levi et al. 2004). SCM crystallizes about integrated business planning that has been espoused by logistics experts, strategists, and operations research practitioners as far back as 1950s. In the last decade, they have been a few changes in business environment that have contributed to the development of SC networks. SCM may be considered as an outcome of globalization and the proliferation of multi-national companies, joint ventures, strategic alliances, and business partnerships; there were found to be significant cannot success factors, following the earlier "Just-In-Time", "Lean Management", and "Agile Manufacturing" practices. SC constitutes a special case of the value chain, a term first introduced by Porter (1985), for those companies that produce and distribute physical products. Competitive advantage accrues to those companies that control their value chain costs better than their competitors. An alternative way to gain competitive advantage is by differentiating their products by providing some combination of superior quality, customer service, product variety, unique market present, and so on (Shapiro 2007).

In today's rapidly changing economic and political conditions, global corpora-
tions face a continuous challenge to constantly evaluate and optimally configure
their SC operations to achieve key performance indices, either it is profitability, cost
reduction, or customer service (Tsiakis and Papageorgiou 2008). The Process
Systems Engineering (PSE) community is performing a key role in extending
system boundaries from chemical process systems to business process systems. PSE
has traditionally been concerned with the understanding and development of systematic procedures for the design, control, and operation of chemical process
systems (Sargent 1991). The scope of PSE can be broadened by making use of the
concept of the "chemical supply chain" (see Fig. 1.2). According to Grossmann and
Westerberg (2000), PSE may be defined as the field of study that is concerned with
the improvement of decision-making processes for the creation and operation of the
"chemical supply chain."

## 1.2.2   Planning and Scheduling

There are three SC decision-making levels: the operational, the tactical, and the
strategic (Shapiro 2007). More specifically, the operational level deals with
short-term scheduling problems, the tactical level involves medium-term planning
decisions, and the strategic level corresponds to the SC design problem. This book
is primarily focused on scheduling and planning problems. However, a contribution
on SC design has been also realized (see Appendix A).

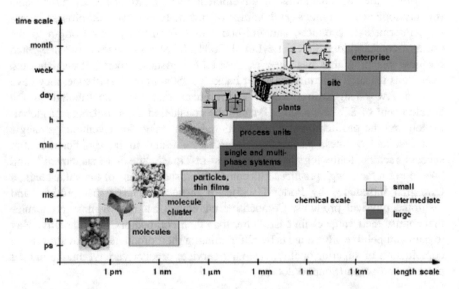

**Fig. 1.2** The "chemical supply chain" (Marquardt et al. 2000)

Kreipl and Pinedo (2004) presented a thorough overview in production planning and scheduling. According to this, a *production planning model* typically optimizes several consecutive stages in an SC (i.e., a multi-echelon model), with each stage having one or more facilities. Such a model is designed to allocate the production of the different products to the various facilities in each time period, while taking into account inventory holding costs and transportation costs. A planning model may make a distinction between different product families, but usually does not make a distinction between different products within a family. It may determine the optimal production run (or, equivalently, the batch size or lot size) of a given product family when a decision has been made to produce such a family at a given facility. If there are multiple families produced at the same facility, then there may be setup costs and setup times. The optimal production run of a product family is a function of the trade-off between the setup cost and/or setup time and the inventory cost. Generally speaking, the main objectives in planning involve inventory costs, transportation costs, tardiness costs, and the major setup costs. However, in a medium-term planning model, it is typically not customary to take the sequence dependency of setup times and setup costs into account. The sequence dependency of setups is difficult to incorporate in an integer programming model because it can increase significantly the complexity of the formulation.

A detailed *scheduling model* is typically concerned with a single facility. Such a model usually takes more detailed information into account than a planning model. It is typically assumed that there are a given number of jobs and each one has its own parameters, including sequence-dependent setup times and sequence-dependent setup costs. The jobs have to be scheduled in such a way that one or more objectives are optimized (e.g., minimization of makespan, weighted lateness, or total setup costs).

Planning models differ from scheduling models in a number of ways. First, planning models often cover multiple stages and optimize over a medium-term horizon (e.g., weeks or months), whereas scheduling models are usually designed for a single facility and optimize over a short-term horizon (e.g., hours or days). Second, planning models use more aggregate information, whereas scheduling models use more detailed information. Third, the objective to be minimized in a planning model is typically a total cost objective and the unit in which this is measured is a monetary unit; the objective to be minimized in a scheduling model is typically a function of the completion times of the jobs and the unit in which this is measured is often a time unit. Nevertheless, even though there are fundamental differences between these two types of models, they often have to be incorporated into a single framework, share information, and interact extensively with one another (Kreipl and Pinedo 2004).

# References

CEFIC (2017) Facts and figures: the European chemical industry in a worldwide perspective. Technical report. Conseil Européen des Fédérations de l'Industrie Chimique

CSCMP (2018). http://cscmp.org. Technical report quarterly. Council of Supply Chain Management Professionals

Grossmann IE, Westerberg AW (2000) Research challenges in process systems engineering. AIChE J 46(9):1700–1703

Kreipl S, Pinedo M (2004) Planning and scheduling in supply chains: an overview of issues in practice. Prod Oper Manage 213(1):77–92

Marquardt W, Wedel LV, Bayer B (2000) Perspectives on lifecycle process modeling. Proc AIChE Symp Ser 96(323):192–214

Min H, Zhou G (2002) Supply chain modeling: past, present and future. Comput Ind Eng 43(1–2):231–249

Porter ME (1985) Competitive advantage: creating and sustaining superior performance. The Free Press, New York

Sargent RWH (1991) What is chemical engineering? CAST Newslett 14(1):9–11

Shapiro JF (2007) Modeling the supply chain. International student edition, Duxbury, New York

Simchi-Levi D, Kaminsky P, Simchi-Levi E (2004) Managing the supply chain: the definitive guide for the business professional. McGraw-Hill, New York

Tsiakis P, Papageorgiou LG (2008) Optimal production allocation and distribution supply chain networks. Int J Prod Econ 111(2):468–483

# Chapter 2
# State of the Art

## 2.1 Introduction

Planning and scheduling techniques have been employed in the process industries since the early 1940s (Shobrys 2001), even before the first computers appeared. Indeed, the oil industry was one of the early adopters of planning methods based on the formulation and solution of linear programming models, with applications already reported in the late 1950s. But, it is since the early 1980s, in particular that the theme of production planning and scheduling for the process industries has received significant attention. Initially, from the early 1980s to the early 2000s, this was due to the resurgence in interest in flexible processing either as a means of ensuring responsiveness or adapting to the trends in process industries toward lower volume, higher added-value materials in the developed economies (Reklaitis 1982). More recently, the topic has received a new impetus as enterprises attempt to optimize their overall Supply Chain (SC) in response to competitive pressures or to take advantage of recent relaxations in restrictions on global trade. The planning and/or scheduling problem at a single site is usually concerned with meeting fairly specific production requirements. Customer orders, stock imperatives or higher level SC or long-term planning would usually set these. Then, the planning/scheduling activity is concerned with the allocation over time of scarce resources between competing activities to meet these requirements in an efficient fashion.

The objective in production planning is to determine the production and inventory levels that will allow fulfilling given customer demand at the minimum cost (including processing, holding and backlog, and switchover costs) subject to (typically aggregate) production capacity constraints. Thus, a production planning solution consists of production targets and inventory levels over a number of periods into which the planning horizon under consideration is partitioned. On the other hand, the objective in scheduling is the allocation of limited resources (e.g., equipment units, utilities, and manpower) to competing tasks and the sequencing and timing of tasks on units, given a set of production targets and subject to detailed

© Springer Nature Switzerland AG 2019
G. M. Kopanos and L. Puigjaner, *Solving Large-Scale Production Scheduling and Planning in the Process Industries*,
https://doi.org/10.1007/978-3-030-01183-3_2

production constraints. The production scheduling component is of vital importance as it is the layer which translates the economic imperatives of the plan into a sequence of actions to be executed on the plant, so as to deliver the optimized economic performance predicted by the higher level plan. Clearly, the two problems are interdependent since the solution of production planning (production targets) is input to scheduling, and the production capacity constraints in production planning depend on the scheduling solution.

The key components of the planning and/or scheduling problem are *resources*, *tasks*, and *time*. The resources need not be limited to processing equipment items, but may include material storage equipment, transportation equipment (intra- and interplant), operators, utilities (e.g., steam, electricity, cooling water), and auxiliary devices and so on. The tasks typically comprise processing operations (e.g., reaction, separation, blending, and packing) as well as other activities that change the nature of materials and other resources such as transportation, quality control, cleaning, changeovers, etc. There are both external and internal elements of the time component. The external element arises out of the need to coordinate manufacturing and inventory with expected product lifting's or demands, as well as scheduled raw material receipts and even service outages. The internal element relates to executing the tasks in an appropriate sequence and at right times, taking into account of the external time events and resource availabilities. Overall, this arrangement of tasks over time and the assignment of appropriate resources to the tasks in a resource-constrained framework must be performed in an efficient fashion, which implies the optimization, as far as possible, of some objective. Typical objectives include the minimization of cost or maximization of profit, maximization of customer satisfaction, minimization of deviation from target performance, etc.

## 2.2 Types of Production Processes

In this section, the major production process patterns found in the process industries are presented. The production process can be classified as *continuous*, *semicontinuous*, and *batch*. A brief description of these processes follows.

### 2.2.1 Continuous Processes

In the continuous processing mode, units are continuously fed and yield a constant product flow. For mass production of similar products, continuous processes can achieve higher consistent product quality, taking advantage of the economies of scale and reducing manufacturing costs and waste. Processes from the petrochemical industry are usually good examples of continuous production processes. The most important difference between batch production and continuous production are that any changes in the product's properties such as color, dimensions, or quality needs to be done online. And whenever it is affected, the results can be seen only after a fixed period that can extend from a few hours to days. Moreover, maintenance in case of continuous process plants calls for online maintenance that requires very high alertness and quick response times from dedicated technicians.

### 2.2.2   Semicontinuous Processes

Semicontinuous processing offers a more customized operation for highly dynamic and uncertain environments. Semicontinuous operations are characterized by their processing rate, running continuously with periodic startups and shutdowns for frequent product transition. The processing times of semicontinuous processes are relatively long periods of time called campaigns, each dedicated to the production of a single product. Typical campaign lengths range from a few hours to several days. Most process plants in the process industry combine continuous operations and batch processes in their product processing routes thus working in semicontinuous mode, since production is more flexible and equipment can be more efficiently utilized.

### 2.2.3   Batch Processes

The primary characteristic of the batch production process is that all components are completed at a workstation before they move to the next one. Batch production is popular in the manufacture of pharmaceutical ingredients, inks, paints, adhesives, and a plethora of contemporary commodities. Batch production is useful for a factory that makes seasonal items or products for which it is difficult to forecast demand. There are several advantages of batch production; for instance, it can reduce initial capital outlay because a single production line can be used to produce several products. Batch production can be useful for small businesses that cannot afford to run continuous production lines. It is worth mentioning that companies may use batch production as a trial run.

Despite the fact that batch processing has been traditionally associated with specialty chemicals and products of high-added value (e.g., pharmaceuticals), the demand patterns can be so unpredictable that profitability may only be achieved by taking full advantage of the inherent flexibility of a batch production facility. Therefore, in order to reduce the risk of new investment, batch plants are preferred as an adequate and flexible answer to the variability in the supply of raw materials, the manufacturing of diverse products and the instability of product demands.

At this point, it is worth mentioning that processing times constitute one of the major differences between scheduling of batch and continuous processes. On the one hand, in batch plants, the processing times are typically fixed and known a priori. Moreover, the production amount depends on the capacity of the batch processing unit. On the other hand, in continuous plants, the processing times are a function of unit-dependent processing rates, final product demand, and storage limitations. Additionally, in continuous plants, the production amount is available continuously while it is being produced, unlike in batch plants, where the produced quantity is available only after the end time of the batch that is being processed.

**Batch Process Types**

According to Rippin (1993), traditional batch (and therefore semicontinuous) production facilities can be classified into *multiproduct* and *multipurpose*.

In multiproduct plants, each product has the same processing network. That means that each product requires the same sequence of processing tasks (often known as stages) and thus, there is only one way to produce a specific product; although some products may skip some task in the sequence. Due to the historic association between the work on batch plant scheduling and that on discrete parts manufacturing, these plants are sometimes called flow shops in the Operational Research (OR) literature. In multipurpose plants, the products are manufactured via different processing networks, and there may be more than one way in which to manufacture the same product. In general, a number of products undergo manufacture at any given time. Flow patterns are not straight lines, as in the multiproduct case, and some units may be used to perform nonconsecutive operations for the same product. Multipurpose plants result in more flexible operation, which can be optimized to decrease equipment idle time to more efficiently utilize critical equipment units. In the OR literature, multipurpose plants are sometimes referred to as job shops. It should be emphasized that these batch processes do not fall under the usual flow shop or job shop in the OR, because the number of jobs (tasks) to be executed is not known a priori, processing times can depend on job size, and jobs are linked to each other through material balance constraints and intermediate storage requirements.

It is worth mentioning that *pipeless plants* have been discussed in the literature (Takahashi and Fujii 1991) as potential alternatives to traditional batch plants. Their main distinguishing feature is that material is transported from one processing stage to another in transferable vessels. Processing takes place at a number of processing stations, and normally the same vessels used for transferring material also hold the material while it is being processed at each station. When necessary, cleaning of the vessels takes place at specialized cleaning stations. The elimination of fixed piping networks for material transfer enables pipeless plants to be considerably more flexible than their conventional counterparts. Nowadays, pipeless plants are still scarce in the process industries and adapted to every particular case. For this reason such plants are not further discussed in this book.

**Intermediates Storage Policies**

Storage for intermediate products plays a significant role in exploiting the inherent flexibility of the batch process. During the operation of batch plants, several material storage policies may be implemented. A list and a short description of the main storage tactics follows.

- *Unlimited Intermediate Storage* (*UIS*)

  This case corresponds to the unrestricted storage case. The material is stable, and one or more dedicated storage vessels are available, the total capacity of which is unlimited. UIS is considered as the best-case scenario, an upper limiting bound for all other solutions, since the best attainable solution with the shortest production times is obtained using this policy.
- *No Intermediate Storage* (*NIS*)

  In this case, the material is stable and there are no storage tanks available for intermediate materials; however, materials can be stored temporarily inside the processing unit, waiting to be transferred to the next unit once it has been empty (this is also true for UIS policy).
- *Zero Wait* (*ZW*)

  This policy restricts the NIS policy by avoiding the alternative use of processing units as storage facilities for intermediate materials. This policy is usually used in cases where the materials are unstable products that must be transferred to the next processing unit immediately after completion. This is the most restrictive policy and constitutes a lower limiting bound.
- *Finite Intermediate Storage* (*FIS*)

  Limited storage capacity is available in terms of the number of storage units, their capacities, and connections between processing units and tanks. The material is stable, and one or more dedicated storage vessels are available, all of which may be subject to optimization.
- *Shared Intermediate Storage* (*SIS*)

  The material is stable, and may be stored in one or more storage vessels that may also be used to store other materials (though not at the same time).

## 2.3 Optimization Methods

All, but the most trivial scheduling problems, belong to the class of NP-hard problems, thus there are no known solution algorithms that are of polynomial complexity in the problem size. This has posed a great challenge to the research community, and a large body of work has arisen aiming to develop either tailored algorithms for specific problem instances or efficient general purpose methods.

### 2.3.1 Mathematical Programming Approaches

The application of mathematical programming approaches implies the development of a mathematical framework and the use of an optimization algorithm. Most mathematical approaches aim to develop models that are of a standard form (from linear programming models for refinery planning to mixed integer nonlinear programming models for multipurpose batch plant scheduling). These models are then usually solved by commercial software or specialized algorithms that take account of the problem structure (for more details, see Chap. 3).

The main decision variables of the mathematical models usually include some or all of the following:

- Selection of resources (e.g., units, utilities) to execute tasks at the appropriate times;
- Sequence of tasks;
- Timing of tasks;
- Amounts processed in each task; and
- Inventory levels of all materials over time.

The Boolean nature of some of the decisions (e.g., sequencing and resource selection) implies the utilization of binary variables. All variables values will be subject to some or all of the following constraints:

- non-preemptive processing once started (i.e., processing activities must proceed until completion);
- resource constraints at anytime (i.e., the utilization of a resource must not exceed its availability);
- material balances;
- capacity constraints for processing and storage; and
- full demand satisfaction for orders by their due dates (if backlogs are not allowed).

The mathematical programming modeling of scheduling and planning in process industries is focused on four key elements: (i) the time representation, (ii) the event representation, (iii) the material balance approach, and (iv) the objective function.

**Time Representation**
The representation of the time horizon is an essential feature of mathematical programming approaches, because processing tasks interact through the use of shared resources and, therefore, the discontinuities in the overall resource utilization profiles must be tracked over time, in order to ensure feasibility by not exceeding the available resource capacities. The complexity arises from the fact that these discontinuities are functions of any schedule proposed and are not known in advance. The three-time representation approaches are:

- *Discrete-time*: the horizon is divided a priori possibly into a number of equally spaced intervals so that any event that introduces such discontinuities (e.g., the starting of a task or a due date for an order) can only take place at an interval boundary. This implies a relatively fine division of the time grid, so as to capture all the possible event times, and in the solution to the problem, it is likely that many grid points will not actually exhibit resource utilization discontinuities.
- *Continuous-time*: the horizon is divided into fewer intervals, the spacing of which will be determined as part of the solution to the problem (i.e., the time

horizon is partitioned as part of the optimization). The number of intervals will correspond more closely to the number of resource utilization discontinuities in the solution.

- *Mixed-time*: the time grid is fixed but the durations of the tasks are variable.

As discussed in Méndez et al. (2006), in discrete-time models, constraints have only to be monitored at certain known time points, a fact that reduces the problem complexity and makes the model structure simpler and easier to solve, especially when resource and inventory limitations are considered. On the other hand, this type of problem simplification has two main drawbacks. First, the size of the mathematical model as well as its computational efficiency strongly depends on the number of time intervals postulated, which is defined as a function of the problem data and the desired accuracy of the solution. Second, suboptimal or even infeasible schedules may be generated because of the reduction of the domain of the timing decisions. In order to overcome the previous limitations and generate data-independent models, a continuous-time representation may be employed. In typical continuous-time formulations, a variable time handling allows obtaining a significant reduction of the number of variables and at the same time, more flexible solutions, in terms of time, can be generated. However, because of the modeling of variable processing times, resource and inventory limitations usually need the definition of more complicated constraints involving many big-M terms, which tends to increase the model complexity and the integrality gap. Discrete-time formulations, despite being a simplified version of the original scheduling problem, have proven to be efficient, adaptable and convenient for some industrial applications, especially in those cases where a reasonable number of intervals are sufficient to obtain the desired problem representation. However, discrete-time models suffer from a number of drawbacks: (i) the discretization interval must be fine enough to capture all significant events, a fact that may result in large model sizes, (ii) it is difficult to model operations where the processing time is dependent on the batch size, and (iii) the modeling of continuous and semicontinuous operations must be approximated, and minimum run lengths give rise to complicated constraints. Concluding, the appropriate selection of the time representation mainly depends on: the production process, the resource limitations, and the objective function of the scheduling and/or planning problem under consideration.

**Material Balance**

Depending on the handling of batches and batch sizes, scheduling formulations can be broadly classified into *network-based formulations* for general processes and *batch-based formulations* for sequential processes. The first category refers to monolithic approaches, which simultaneously deal with the lot-sizing and scheduling problems. These methods are able to deal with arbitrary network processes involving complex product recipes. However, their generality usually implies large model sizes and consequently, their application is currently restricted to processes involving a small number of processing tasks and rather short scheduling horizons. Batch-based formulations are used for single-stage, multistage and multipurpose processes where batches are processed sequentially and where

16                                                         2 State of the Art

batch splitting/mixing are not allowed and there are no recycle streams. In these approaches, the number and the size of batches are known in advance. In other words, the lot-sizing (or batching) problem has already been solved. The batching problem converts the demand of products into individual batches aiming at optimizing some criterion. Afterward, the available manufacturing resources are allocated to the batches over time. This approximate two-stage approach, widely used in industry, can address much larger practical problems than monolithic methods, especially those involving a quite large number of batch tasks related to different intermediates or final products (Méndez et al. 2006). Network-based formulations typically employ the State-Task Network or the Resource-Task Network process representation in order to formally represent the problem. A description of these process representations follows.

**State-Task Network (STN).** Kondili et al. (1988) introduced the STN process representation, by presenting a discrete-time Mixed Integer Programming (MIP) model. The STN is a directed graph that consists of three key elements: (i) *state nodes* representing feeds, intermediates and final products, (ii) *task nodes* representing the process operations which transform material from one or more input states into one or more output states, and (iii) *arcs* that link states and tasks indicating the flow of materials. Circles and rectangles denote state and task nodes, respectively (see Fig. 2.1).

The three main advantages of the STN representation are that:

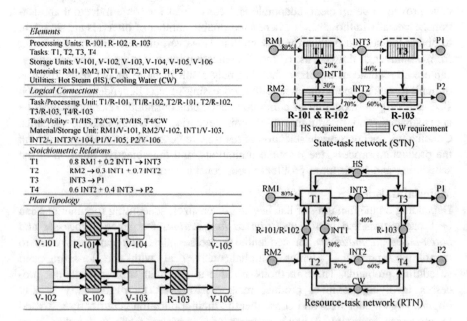

**Fig. 2.1** STN and RTN process representation examples (Gimenez et al. 2009a)

(i)   It distinguishes the process operations from the resources that may be used to execute them, and therefore provides a conceptual platform from which to relax the unique assignment assumption and optimize unit-to-task allocation.
(ii)  It avoids the use of task precedence relations that become very complicated in multipurpose plants. A task can be scheduled to begin if its input materials are available in the correct amounts and other resources (e.g., processing equipment and utilities) are also available, regardless of the plant history.
(iii) It provides a means of describing very general process recipes, involving batch splitting and mixing, material recycles, and intermediate storage.

As argued by Pantelides (1994) STN, despite its advantages, suffers from a number of drawbacks:

(i)   The model of plant operation is somewhat restricted, since each task is assumed to use exactly one major item of equipment during its operation.
(ii)  Tasks are always assumed to be processing activities that change material states, therefore changeovers or transportation activities have to be treated as special cases.
(iii) Each item of equipment is treated as a distinct entity, a fact that introduces solution degeneracy if multiple equivalent items exist.
(iv)  Different resources (e.g., materials, units, and utilities) are treated differently, giving rise to many different types of constraints, each of which must be formulated carefully to avoid unnecessarily increasing the integrality gap.

**Resource-Task Network (RTN).** Pantelides (1994) proposed the RTN process representation, which is based on a uniform description of all resources, as a more general case of the STN representation. In the RTN representation, a task is assumed only to consume and produce resources, in contrast to the STN, where a task consumes and produces materials while using equipment and utilities during its execution. Processing items are treated as though consumed at the beginning of a task and produced at the end. In the RTN representation, circles represent not only states but also other resources required in the batch process such as processing units and vessels (see Fig. 2.1). A special feature of the RTN is that processing equipment in different conditions (e.g., "clean" or "dirty") can be treated as different resources, with diverse activities (e.g., "processing" or "cleaning") consuming and generating them, thus achieving a simpler representation of changeovers. The RTN main advantage over the STN representation lies in its conceptual simplicity and its direct applicability to a large number of complex process scheduling problems. In scheduling problems involving identical equipments, RTN-based formulations overwhelm STN-based models, since they introduce just a single binary variable instead of the multiple variables used by the STN-based models. In a few words, RTN-based models reduce the batch scheduling problem to a simple resource balance problem carried out in each predefined time period (Méndez et al. 2006). The ability to capture additional problem features in a straightforward fashion made the RTN representation a promising framework for future research.

**Event Representation**
In addition to the time representation and material balances, scheduling models are based on different concepts or basic ideas that arrange the events of the schedule over time with the main purpose of guaranteeing that the maximum capacity of the shared resources is never exceeded (Méndez et al. 2006). In Fig. 2.2 are illustrated by the five basic event representation concepts. Namely, they are (a) *global time intervals*, (b) *global time points*, (c) *unit-specific time events*, (d) *time slots*, and (e) *precedence-based*.

Global time intervals are used in discrete-time models. In this event representation, fixed time grids are predefined and the tasks are forced to begin and finish exactly at a point of the grid (see Fig. 2.2a). Consequently, the original scheduling problem is reduced to an allocation problem where the main model decisions denote the assignment of the time interval at which every task begins. The contributions of Kondili et al. (1988), Kondili et al. (1993), and Shah et al. (1993a) are based on this concept.

To continue with, in contrast with discrete-time models, there is a variety of event representations in continuous-time domain formulations. More specifically, network-based models for general processes use global time points or unit-specific time events, while batch-based formulations employ time slots or precedence-based relationships. In the global time point's concept, the timing of time intervals is treated as a new model variable. Thus, a common and a variable time grid are defined for all shared resources while the starting and the finishing times are linked to specific time points through key discrete variables (see Fig. 2.2b). Some of the most important works that use this concept are the ones of Castro et al. (2001), where an RTN representation is used, and Maravelias and Grossmann (2003). The unit-specific time events concept defines a different variable time grid for each shared resource, allowing different tasks to start at different moments for the same event point (see Fig. 2.2c). This concept results in more complicated models compared with the global time point's concept, since, due to lack of references

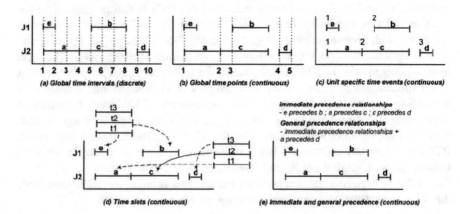

(a) Global time intervals (discrete)     (b) Global time points (continuous)     (c) Unit specific time events (continuous)

(d) Time slots (continuous)               (e) Immediate and general precedence (continuous)

**Fig. 2.2** Types of event representation (Méndez et al. 2006)

points, additional constraints and variables need to be defined for dealing with shared resources. Representative works of the unit-specific time events concept are the ones reported by Ierapetritou and Floudas (1998) and Giannelos and Georgiadis (2002).

The previous event representations are used in network-based formulation for general processes. For sequential processes, time slots and precedence-based formulations have been developed. Indeed some of them have been recently extended to also consider general processes. In the time slots concept, a set of an appropriate number of predefined time intervals for each processing unit with unknown durations is firstly postulated in order to allocate them to the tasks (see Fig. 2.2d). The choice of the number of time slots required represents an important trade-off between optimality and computational performance. The work of Sundaramoorthy and Karimi (2005) has been based on this concept.

At this point, it should be emphasized that in all the aforementioned event representation concepts a predefined number of time points or slots is needed; a fact that may affect the problem optimal solution if the right number of them is not considered. For this reason, alternative approaches that are based on the concept of task precedence have been emerged (see Fig. 2.2e). In these formulations, sequencing binary variables, through big-M constraints, enforcing the sequential use of shared resources are explicitly employed. As a result, sequence-dependent changeovers can be treated in a straightforward manner.

Three precedence concepts have been reported: (i) the immediate precedence, (ii) the unit-specific immediate precedence, and (iii) the general precedence. The immediate precedence concept explores the relation between each pair of consecutive tasks without considering if the orders are assigned or not into the same unit. A representative example of an immediate precedence formulation can be found in Méndez et al. (2000). The unit-specific immediate precedence is based on the immediate precedence concept. The difference is that it takes into account only the immediate precedence of the tasks that are assigned to the same unit. The formulation presented by Cerdá et al. (1997) is a representative example of unit-specific immediate precedence models. The general precedence generalizes the precedence concept by exploring the precedence relations of every task regarding all the remaining tasks and not only the immediate predecessor. This approach results in a lower number of binary variables, and comparing it with the other two approaches, reducing significantly the computational effort on average. The work of Méndez and Cerdá (2003a) is based on the general precedence concept. Concluding, it should be mentioned that a common weakness of precedence-based formulations is that the number of sequencing variables scales in the number of batches to be scheduled, which may result in significant model sizes for real-world applications.

**Objective Function**

Different measures of the quality of the solution can be used for scheduling and planning problems. The selection of the optimization goal directly affects the solution quality as well as the model computational performance. Typical objective functions include the optimization of makespan, weighted lateness, production costs, inventory costs, total cost, revenue, and profit. It should be noted that some objective functions can be very hard to implement for some event representations, requiring additional variables and complex constraints.


## 2.3.2   Alternative Solution Approaches

Although this book has been focused on solution approaches based on mathematical programming techniques, it is important to note that there are other solution methods for dealing with process scheduling and/or planning problems. These methods can be used either as alternative methods, or as methods that can be combined with mathematical programming models. The major alternative solution methods for solving scheduling problems are heuristics, metaheuristics, artificial intelligence, constraint programming, and hybrid methods. In addition to these methods, Process Systems Engineering (PSE) research community developed two elaborate approaches to deal with process scheduling problems: the event operation network representation, and the S-graph representation.

### Heuristics
Most scheduling heuristics, also called dispatching rules, are concerned with formulating rules for determining sequences of activities. They are therefore best suited to processes where the production of a product involves a pre-specified sequence of tasks with fixed batch sizes; in other words variants of multiproduct processes. Often, it is assumed that fixing the front-end product sequence will fix the sequence of activities in the plant. Generally, the processing of a product is broken down into a sequence of jobs that queue for machines, and the rules dictate the priority order of the jobs. Dannenbring (1977), Kuriyan and Reklaitis (1989), and Pinedo (1995) give a good exposition on the kinds of heuristics that may be used for different plant structures. It is worth pointing out that *most of the heuristic methods originated in the discrete manufacturing industries, and might sometimes be expected to perform poorly in process industry contexts*. In process scheduling problems, most of the concerns with these approaches are associated with the divisibility of material in practice, which implies variable batch sizes, and batch splitting and mixing. In fact, the last two activities are becoming increasingly popular as a means of effecting late product differentiation. Some of the most broadly used dispatching rules are First Come First Served (FCFS), Earliest Due Date (EDD), Shortest Processing Time (SPT), Longest Processing Time (LPT), Earliest Release Date (ERD), and Weighted Shortest Processing Time (WSPT). Often, composite dispatching rules, involving a combination of basic rules, can perform significantly better. Besides, a special feature of heuristics is that they can be easily embedded in mathematical models to generate more efficient hybrid approaches for large-scale scheduling problems. An extensive review and a classification of various heuristics can be found in Panwalkar and Iskander (1977) and Blackstone et al. (1982).

There is a lack of works that developed heuristics for process scheduling problems, since it is difficult to devise a series of rules to solve such complex processes. Kudva et al. (1994) addressed the special case of linear multipurpose plants where products flow through the plant in a similar fashion, but potentially using different stages and with no recycling of material. They took account of limited intermediate storage, material receipts at any stage, soft order deadlines, changeover costs and pre-specified equipment maintenance times. A rule-based constructive heuristic was used, which required the maintenance of a status sheet on each unit and material type for each time instance on a discrete-time grid. The algorithm used this status sheet with a sorted list of orders and developed a schedule for each order by backward recursive propagation. The schedule derived depended strongly on the order sequence. Solutions were found to be within acceptable bounds of optimality when compared with those derived through formal optimization procedures. Graells et al. (1996) presented a heuristic strategy for the scheduling of multipurpose batch plants with mixed intermediate storage policies. A decomposition procedure was employed where sub schedules were generated for the production of intermediate materials. Each subschedule consisted of a mini production path determined through a branch-and-cut enumeration of possible unit-to-task allocations. The mini-paths were then combined to form the overall schedule. The overall schedule was checked for feasibility with respect to material balances and storage capacities. Improvements to the schedules may be effected manually through an electronic Gantt chart.

It should be emphasized that the implementation of heuristics to scheduling problems in the process industries is not straightforward. For this reason, most academic research has been directed toward the development of mathematical programming approaches for process scheduling and planning, since these approaches are capable of representing all the complex interactions in such complex processing networks.

**Metaheuristics**

Metaheuristics are often inspired by moves arising in natural phenomena. Metaheuristics optimize a problem by iteratively trying to improve a candidate solution with regard to a given measure of quality. Metaheuristics such as genetic algorithms, graphs theory, simulated annealing, tabu search, particle swarm, and ant colony optimization methods have been widely used in a variety of scheduling problems. These techniques have become popular for optimizing certain types of scheduling problems, however, they also have significant drawbacks such as that they do not provide any guarantee on the quality of the solution obtained, and it is often impossible to tell how far the current solution is from optimality. Furthermore, these methods do not formulate the problem as a mathematical program, since they involve procedural search techniques that in turn require some type of discretization or graph representation, and the violation of constraints is handled through ad hoc penalty functions. For this reason, the use of metaheuristics might be problematic for problems involving general processes, complex inequality constraints and continuous decision variables. In this case, the set of feasible solutions might lack nice properties and it might even be difficult to find a feasible solution (Burkard et al. 1998). Some excellent contributions in this direction can be found in Kirkpatrick et al. (1983), Glover (1990), Ku and Karimi (1991), Xia and Macchietto (1994), Franca et al. (1996), Murakami et al. (1997), Raaymakers and Hoogeveen (2000), Pacciarelli (2002), Cavin et al. (2004), Ruiz and Maroto (2006), Ruiz and Stutzle (2008) and Venditti et al. (2010), among many others. Despite the fact that the aforementioned methods may generate fast and effective solutions for complex problems, they are usually tailor-made and cannot systematically estimate the degree of quality of the solution generated. Moreover, the efficiency of these techniques strongly depends on the proper implementation and fine tuning of parameters since they combine the problem representation and the solution strategy into the same optimization framework.

**Artificial Intelligence Methods**

Artificial Intelligence (AI) is the mimicking of human thought and cognitive processes to solve complex problems automatically. AI uses techniques for writing computer code to represent and manipulate knowledge. There are different techniques that mimic the different ways that people think and reason. The main AI techniques are rule-based methods, agent-based methods, and expert systems. Rule-based methods can be distinguished into case-based reasoning and model-based reasoning techniques. Case-based reasoning is based on previous experiences and patterns of previous experiences while model-based reasoning concentrates on reasoning about a system's behavior from an explicit model of the mechanisms underlying that behavior. Agent-based approaches are software programs that are capable of autonomous, flexible, purposeful and reasoning action in pursuit of one or more goals. They are designed to take timely action in response to external stimulus from their environment on behalf of a human. Expert systems, also known as knowledge-based approaches, encapsulate the specialist knowledge gained from a human expert and apply that knowledge automatically to make decisions. Some interesting implementations of AI technologies into real-world scheduling problems can be found in Zweben and Fox (1994), Sauer and Bruns (1997), and Henning and Cerdá (2000).

**Constraint Programming**

Constraint programming is a programming paradigm that was originally developed to solve feasibility problems (Van Hentenryck 1989, 2002), but it has been extended to solve optimization problems, particularly scheduling problems. Constraint programming is very expressive since continuous, integer, and Boolean variables are permitted; moreover, variables can be indexed by other variables. Furthermore, a number of constructs and global constraints have also been developed to efficiently model and solve specific problems, and constraints need neither be linear nor convex. The solution of constraint programming models is based on performing constraint propagation at each node by reducing the domains of the variables. If an empty domain is found, the node is pruned. Branching is performed whenever a domain of an integer, binary or Boolean variable has more than one element, or when the bounds of the domain of a continuous variable do not lie within a tolerance. Whenever a solution is found, or a domain of a variable is reduced, new constraints are added. The search terminates when no further nodes must be examined. The effectiveness of constraint programming depends on the propagation mechanism behind constraints. Thus, even though many constructs and constraints are available, not all of them have efficient propagation mechanisms. For some problems, such as scheduling, propagation mechanisms have been proven to be very effective. Constraint programming methods have proved to be quite effective in solving certain types of scheduling problems, especially those that involve sequencing and resource constraints. However, they are not always effective for solving more general optimal scheduling problems that involve assignments (Méndez et al. 2006).

Some of the most common propagation rules for scheduling are the "timetable" constraint (Le Pape 1998), the "disjunctive constraint" propagation (Baptiste et al. 2001), the "edge-finding", and the "not-first, not-last" (Baptiste et al. 2001). Finally, Laborie (2003) summarized the main approaches to propagate resource constraints in constraint-based scheduling and identified some of their limitations for using them in an integrated planning and scheduling framework.

**Hybrid Methods**

In this paragraph, some important hybrid solution techniques, applied in process scheduling, based on exact solution methods (i.e., mathematical programming) are presented. It should be emphasized that although small- and medium-size models can be usually solved to optimality by using default values in code parameters, large size problems are generally unmanageable by mathematical formulations. Therefore, in order to make the use of exact methods more attractive in real-world applications, an increasing effort has been oriented toward the development of systematic techniques that allow maintaining the number of decisions at a reasonable level, even for large-scale problems. A reduced search space usually results in manageable model sizes that often guarantee a more stable and predictable optimization model behavior. Furthermore, once the best possible feasible solution has been generated in a short time, optimization-based methods could be employed to gradually enhance a nonoptimal solution with low computational effort. Following this trend, the work of Castro et al. (2009) has been recently emerged as alternative solution strategies to these challenging problems. An apparent drawback of these techniques is that optimality can no longer be assured. Nevertheless, from a practical point of view, guaranteeing global optimality may not be relevant in many industrial scenarios mainly due to the following features: (i) a very short time is just available to generate a solution and send it to the plant floor, (ii) optimality is easily lost because of the highly dynamic nature of industrial environments, (iii) implementing the schedule as such is limited by the real process, and (iv) only a part of the real scheduling goals are generally taken into account in the model since not all scheduling objectives can be quantified. Heuristic model reduction methods, decomposition/aggregation techniques, and improvement optimization-based techniques constitute the principal methods that are embedded in exact mathematical models to face large-scale scheduling problems. A detailed state-of-the-art of these techniques can be found in Méndez et al. (2006). A brief description of the aforementioned methods follows.

- Heuristic model reduction methods usually take into advantage an empirical solution tactic or a particular problem feature and incorporate this knowledge into the mathematical problem representation. As a result, good solutions can be generated in a reasonable time. Simple or combined dispatching rules are usually adopted. The contributions by Pinto and Grossmann (1995), Cerdá et al. (1997), Blömer and Günther (2000) and Méndez et al. (2001) are some representative works of heuristic model reduction methods.
- Approaches based on spatial or temporal decomposition such as the works by Graves (1982), Lázaro and Puigjaner (1988) and Gupta and Maranas (1999), usually rely on Lagrangian decomposition. Aggregation techniques aggregate later time periods within the specified time horizon in order to reduce the dimensionality of the problem, or to aggregate the scheduling problem so that it can be considered as part of a planning problem [refer to the works of Birewar and Grossmann (1990) and Bassett et al. (1997)].
- Improvement optimization-based techniques can be interpreted as a special case of rescheduling where an initial solution is partially adjusted with the only goal of enhancing a particular scheduling criterion. These techniques use the current schedule as the initial point of a procedure that iteratively enhances the existing solution in a systematic manner. The works by Röslof et al. (2001) and Méndez and Cerdá (2003a), which followed this direction, have shown promising results with relatively low computational cost.

**Event Operation Network Representation**

Graells et al. (1998) provided a realistic representation of complex recipes that uses a flexible modeling environment to schedule batch chemical processes. The structure of the process (individual tasks, entire subtrains, or complex structures of manufacturing activities) and the related materials (e.g., raw materials, intermediate, or final products) are characterized by a processing network, which describes the material balance. In the most general case, the activity that is carried out in each process constituted a general activity network. Manufacturing activities are considered at three levels of abstraction: the *process level*, the *stage level*, and the *operation level*. This hierarchical approach considers material states (subject to material balance and precedence constraints) and temporal states (subject to time constraints) at different levels.

At the process level, the process and materials network provides a general description of production structures (such as synthesis and separation processes) and of the materials involved, including intermediates and recycled materials. An explicit material balance is specified for each of the processes in terms of a stoichiometric-like equation that relates raw materials, intermediates and final products (see Fig. 2.3). Each process may represent any kind of activity that is required to transform the input materials into the derived outputs.

The stage level lies between the process level and the detailed description of the activities involved at the operation level. At this level, the block of operations that are executed in the same equipment is described. Hence, at the stage level each

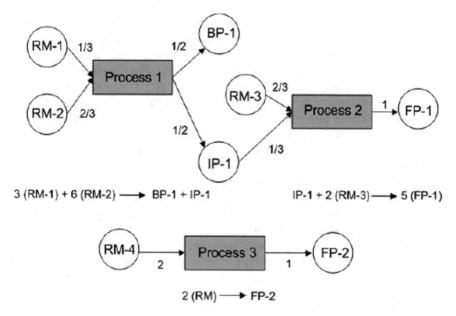

3 (RM-1) + 6 (RM-2) ⟶ BP-1 + IP-1                    IP-1 + 2 (RM-3) ⟶ 5 (FP-1)

2 (RM) ⟶ FP-2

**Fig. 2.3** Process and materials network describing the processing of two products. RM, IP, and FP are raw materials, intermediate products, and final products, respectively

process is split into a set of the blocks (see Fig. 2.4). Each stage involves the following constraints:

- The sequence of operations that are involved requires a set of implicit constraints (links).
- Unit assignment is defined at this level. Thus, for all the operations in the same stage, the same unit assignment must be made.
- A common size factor is attributed to each stage. This size factor summarizes the contribution of all the operations involved.

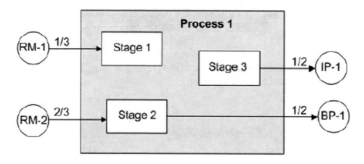

**Fig. 2.4** Stage level. Each stage involves different unit assignments

The operation level contains a detailed description of all the activities considered in the network. Implicit time constraints (links) must also be met at this level. The detailed representation of the structure of activities that define the different processes is called the Event Operation Network (EON). The general utility requirements (e.g., renewable, nonrenewable, storage) are also represented at this level. The EON representation model describes the appropriate timing of process operations. A continuous-time representation of process activities is made using three basic elements: events, operations and links (Puigjaner 1999). Events designate the time instants in which some change occurs. They are represented by nodes in the EON graph, and may be linked to operations or other events. Each event is associated with a time value and a lower bound.

Operations comprise the time intervals between events (see Fig. 2.5). A box linked with solid arrows to its associated nodes represents each operation $m$: initial $NI_m$ and final $NF_m$ nodes. Operations establish the equality links between nodes, in terms of the characteristic properties of each operation: The Operation Time (TOP), and the Waiting Time (TW). The operation time will depend on the amount of materials to be processed, the unit model and the product changeover. The waiting time is the lag time between operations, which is bounded. Finally, precedence constraints are used to establish links between events.

A dashed arrow represents each link $K$ from its node of origin $NO_k$ to its destiny node $ND_k$ and an associated offset time $\Delta T_k$ (see Fig. 2.6).

Despite its simplicity, the EON representation is very general and flexible and it allows complex recipes to be handled (see Fig. 2.7). The corresponding TOP, according to the batch size and material flow rate, also represents transfer operations between production stages. The necessary time overlap of semicontinuous operations with batch units is also considered in this representation by means of appropriate links.

Plant operation can be simulated by means of the EON representation using the following information, which is contained in the process recipe and production structure characteristics:

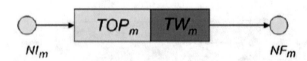

**Fig. 2.5** The time description for operations. TOP = operation time, TW = waiting time $NI_m$ = initial node, and $NF_m$ = final node of operation

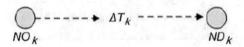

**Fig. 2.6** Event to event link and associated offset time

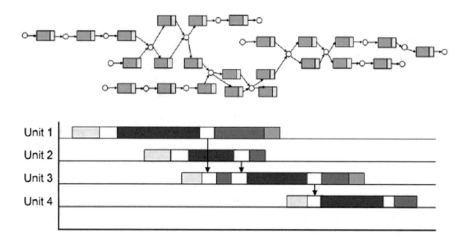

**Fig. 2.7** EON representation of a branched complex recipe. The Gantt chart is given below

- A sequence of production runs or jobs associated with a process or recipe.
- A set of assignments that is associated with each job and consistent with the process.
- A batch size that is associated with each job and is consistent with the process.
- A set of shifting times for all of the operations involved.

These decisions may be generated automatically using diverse procedures to determine an initial feasible solution. Hence, simulation may be executed by solving the corresponding EON to determine the timing of the operations and other resource requirements.

**S-Graph Representation**

Sanmartí et al. (1998, 2002) introduced a graph representation for solving process scheduling problems. This scheduling graph, called S-graph, takes into consideration the specific characteristics of chemical processes in the scheduling. It allows scheduling problems to be formulated using similar graph representations to those used to solve the job shop problem. However, it takes into account the higher complexity of chemical multipurpose batch scheduling.

The master recipes are represented as a directed conjunctive graph, in which the nodes represent the production tasks and the arcs are the precedence relationships among tasks. The number above the arrows represents the task processing times. An additional node is associated with each product: the last task or tasks of the production are connected to the corresponding node by an arc. Thus, for each product, the number of nodes in the graph is the number of tasks in the recipe plus one. For instance, in Fig. 2.8, the sequence of tasks A1 → A2 → A3 is displayed using a graph of this type. The graph consists of four nodes instead of three. The fourth node is required to represent the end of the last task A3, since the completion of task A1 coincides with the start of task A2, and the completion of A2 with the start of A3.

**Fig. 2.8** S-graph
representation of a sequence
of tasks

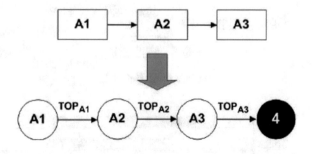

Complex recipes can be represented in this way. Figure 2.9a illustrates the conventional representation of the recipes of three products, in which two intermediates are produced, mixed, and further processed in the production of the first product A. Figure 2.9b shows the graph representation of the recipes given in Fig. 2.9a, where Ei denotes the set of equipment units that can perform the task represented by node i (e.g., E1 = {1, 2}). Furthermore, an additional node (the product node) is introduced for each product. In this representation, the value assigned to an arc expresses a lower bound for the difference between the start times of the two-related tasks. The processing time of a task may vary for different equipment units. In this case, the weight of the arc is the minimum of the processing times of the plausible equipment units. The S-graph uses the recipe representation described above to find a single solution for a scheduling problem. There is one schedule graph for each feasible schedule of the problem. The S-graph $G'(N, 1, A_2)$ is called a schedule graph of the recipe graph $G(N, 1, \emptyset)$, if all the tasks represented in the recipe graph have been scheduled by taking into account equipment-task assignments. The schedule graph of the optimal schedule can be effectively generated by an appropriate search strategy, which enables early detection of infeasible schedules (Sanmartí et al. 1998, 2002).

Initially, S-graph was only applied to problems considering minimization the makespan. The problem was solved using a branch and bound and an efficient graph algorithm to evaluate the makespan. Afterward, S-graph capabilities were further extended so that there is now an effective search algorithm for determining schedules that optimize throughput, revenue, or profit over a predefined time horizon in multipurpose batch plants (Majozi and Friedler 2006), and taking into account the aspects of uncertainty (Laínez et al. 2010a).

## 2.4  Uncertainty

Process industries are dynamic in nature and, therefore, different kinds of unexpected events occur quite frequently. Unexpected disturbances affect the nominal operating conditions and the, now out-of-day, schedule of the production facility. A lack of appropriate procedures for tackling disruptions caused by uncertain

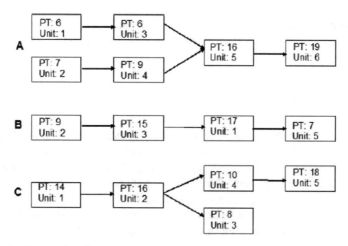

(a) Conventional representation of master recipes of three products.

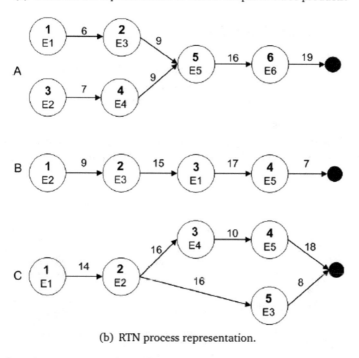

(b) RTN process representation.

**Fig. 2.9**  S-graph representation of recipes shown

events yields into significant performance deterioration. Despite the fact that the study of uncertainty is out of the scope of the current book (however, in Chap. 6 some unexpected scenarios are considered, and efficiently tackled online), in this section a brief description of the major methods for managing uncertainty is given.

More details regarding optimization under uncertainty in process industries can be found in the neatly written state-of-the-art review by Sahinidis (2004).

## 2.4.1   Uncertainty Sources

A taxonomy of the main sources of uncertainty in each SC decision-making level is given in Fig. 2.10. It is important to notice that most sources of uncertainty do not fit totally within one of these categories, but the boundaries are somehow diffuse. Besides, because of the interactions between the different levels of decision-making, uncertainties from one level may affect decisions made in other levels. For instance, variable demands do not only alter tactical planning decisions, but also the process scheduling in the operational level (Bonfill 2006).

## 2.4.2   Managing Uncertainty

The methods for managing uncertainty can be mainly distinguished into *proactive* (*offline*) and *reactive* (*online*) approaches. Figure 2.11 depicts the different techniques for each approach of dealing with uncertainty.

**Proactive Approaches**
Proactive methods, and mainly proactive scheduling approaches, can be viewed as sub-optimization strategies that provide visibility for future actions to achieve a

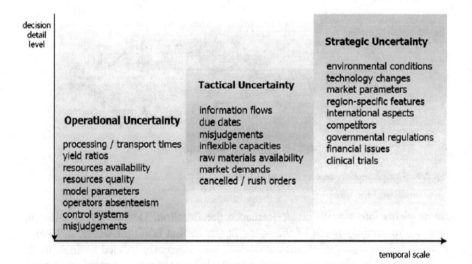

**Fig. 2.10**   A taxonomy of uncertainty sources (Bonfill 2006)

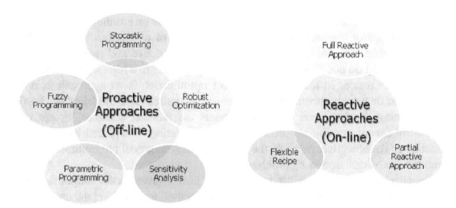

**Fig. 2.11** Methods for managing uncertainty

greater system's performance. If the uncertainty occurs as predicted, the loss of opportunities and reschedule requirements are reduced, whereas the full force of the perturbation affects the expected results if the uncertainty is neglected (Aytug et al. 2005).

**Stochastic-based approaches**. This is the most commonly used approach in the literature. The original deterministic mathematical model is transformed into a stochastic model treating the uncertainties as random variables. Stochastic approaches are mainly divided into the following categories:

- two-stage, where variables are separated to first stage (or here-and-now) decisions and to second stage (or wait-and-see) decisions, or multistage stochastic programming:
  - (i)   scenario-based,
  - (ii)  probabilistic distribution.
- chance constraint programming-based approach.

**Fuzzy programming methods**. The main difference between stochastic and fuzzy optimization approaches is in the way uncertainty is modeled.

Here, fuzzy programming considers random parameters as fuzzy numbers and constraints are treated as fuzzy sets. Some constraint violation is allowed and the degree of satisfaction of a constraint is defined as the membership function of the constraint. Objective functions in fuzzy mathematical programming are treated as constraints with the lower and upper bounds of these constraints defining the decision makers expectations. Fuzzy logic and probability are different ways of expressing uncertainty. While both fuzzy logic and probability theory can be used to represent subjective belief, fuzzy set theory uses the concept of fuzzy set membership (i.e., how much a variable is in a set), and probability theory uses the concept of subjective probability (i.e., how probable do I think that a variable is in a set).

**Robust optimization methods**. These methods focus on building the preventive scheduling and/or planning to minimize the effects of disruptions on the performance measure. They also try to ensure that the predictive and realized schedule and/or planning do not differ drastically, while maintaining a high level of schedule and/or performance. In mathematics, robust optimization is an approach in optimization to deal with uncertainty. It is similar to the recourse model of stochastic programming, in that some of the parameters are random variables, except that feasibility for all possible realizations (called scenarios) is replaced by a penalty function in the objective. As such, the approach integrates goal programming with a scenario-based description of problem data.

**Sensitivity analysis**. It is used to ascertain how a given model output depends upon the input parameters. This is an important method for checking the quality of a given model, as well as a powerful tool for checking the robustness and reliability of any solution (Li and Ierapetritou 2008). Sensitivity analysis determines, on individual parameters of the model, the range in which the solution remains optimal, provided all other parameters are fixed at their given values. Although valuable knowledge can be obtained, sensitivity analysis is usually considered as a post-optimization approach that does not provide any mechanism to control and improve the robustness of a given proposed solution (Mulvey et al. 1995).

**Parametric programming methods**. Parametric optimization serves as an analytic tool in process synthesis under uncertainty mapping the uncertainties in the definition of the synthesis problem to optimal design alternatives. From this point of view, it is the exact mathematical solution of the uncertainty problem. Parametric programming can be used into a model predictive control framework for online optimization via offline (parametric) techniques. Parametric programming techniques have been developed and proposed as a means of reducing computational effort associated to optimization problems regarding uncertainty. To address such problems by using the aforementioned techniques, it is obtained a complete map of all the optimal solutions. As a result, as the operating conditions fluctuate, one does not have to re-optimize for the new set of conditions since the optimal solution is already available as a function of parameters (or the new set of conditions) (Pistikopoulos et al. 2002).

### Reactive Approaches

Reactive methods deal with uncertainty after the occurrence of the unexpected events. Since they tackle unforeseen events online, they should be fast enough, computationally speaking, in order to be applicable to the industrial environment. Despite the fact that the study of uncertainty is out of the scope of this book, a contribution on reactive scheduling, where the importance of considering rescheduling costs is highlighted, has been also realized (see Appendix A).

**Full reactive approaches**. They make decisions dynamically when some event occurs by permitting full alterations of the current schedule/planning. It is the most computationally expensive approach. Additionally, its proposed solution is usually very difficult to be applied to the real industrial scenario because of the large number of modifications that these approaches propose.

**Partial reactive approaches**. They are based on the modification of the pre-dictive schedule and/or planning to update the decisions according to the actual situation by allowing a limited number of modifications. Computational effort is moderated. However, the problem may be over-restricted, thus disregarding potential optimal solutions.

**Flexible recipe**. Process operating conditions are modified in order to adjust the processing times so as to return to the original requirements. The major drawback of this procedure is that there may be little flexibility for the modification of these conditions to guarantee the quality of the products.

## 2.5   Literature Review

In this section, a literature review in the research field of scheduling and/or planning problems by mathematical programming approaches in the process industry is presented. A separate literature review for the food process industry is also included because it corresponds to an emerging, promising and challenging field of research that has received little attention despite its significant role in contemporary markets.

### 2.5.1   Process Industry

The literature review of the most important mathematical approaches addressing scheduling and/or planning problems in the process industry have been classified into discrete-time and continuous-time models, and multisite and resource-constrained production.

**Discrete-Time Models**

Kondili et al. (1988, 1993) introduced the STN process representation and they presented a discrete-time MIP model. This model was based on the definition of binary variables that indicate whether tasks start in specific units at the beginning of each time interval, together with associated continuous batch sizes. Other key variables were the amount of material in each state held in dedicated storage over each time interval, and the amount of each utility required for processing tasks over each time interval. The key constraints were related to resources (i.e., processing units and utilities), material balances and capacity. The use of a discrete-time grid captured all the plant resource utilization in a straightforward manner; disconti-nuities in these were forced to occur at the predefined interval boundaries. However, this approach was hindered in its ability to handle large problems by the weakness of the allocation constraints and the general limitations of discrete-time approaches such as the need for relatively large numbers of grid points to represent activities with significantly different durations.

The emergence of the STN representation formed the basis of many other works aiming to take into advantage of the representational capabilities of the formulation while enhancing its computational performance. More specifically, Sahinidis and Grossmann (1991) disaggregated the allocation constraints and exploited the embedded lot-sizing nature of the model where relatively small demands were distributed throughout the scheduling horizon. The computational performance of the model was improved, despite the larger nature of the disaggregated model. In addition, the formulation of Sahinidis and Grossmann (1991) had a much smaller integrality gap than the original STN model. Afterward, Shah et al. (1993a) modified the allocation constraints even further to generate the smallest possible integrality gap for this type of formulations. They also devised a tailored branch-and-bound solution procedure that utilizes a much smaller LP relaxation and solution processing to improve integrality at each node. The same authors Shah et al. (1993b) considered the extension to cyclic scheduling, where the same schedule was repeated at a frequency to be determined as part of the optimization. Papageorgiou and Pantelides (1996a, 1996b) extended this to cover the case of multiple campaigns, each with a cyclic schedule to be determined.

Elkamel (1993) also proposed a number of measures to improve the performance of the STN-based discrete-time scheduling model. A heuristic decomposition method was proposed, which solves separate scheduling problems for parts of the overall scheduling problem. The decomposition may be based on the resources (longitudinal decomposition) or on time (axial decomposition). In the former, the recipes and suitable equipment for each task were examined for the possible for-mation of unique task-unit subgroups that can be scheduled separately. Axial decomposition was based on grouping products by due dates and decomposing the horizon into a series of smaller time periods, each concerned with the satisfaction of demands falling due within it. In addition, a perturbation heuristic was described, which actually was a form of local search around the relaxation.

Yee and Shah (1997, 1998) also considered various manipulations to improve the performance of general discrete-time scheduling models. An important feature of their work was the variable elimination, since they recognized that in such models, only about 5–15% of the variables reflecting task-to-unit allocations were active at the integer solution, and it would be beneficial to identify as far as possible inactive variables prior to solution. For this reason, they proposed an LP-based heuristic, flexibility and sequence reduction technique, and a formal branch-and-price method. They also recognized that some problem instances resulted in poor relaxations and propose valid inequalities and a disaggregation procedure similar to that of Sahinidis and Grossmann (1991) for particular data instances.

Gooding (1994) considers a special case of the problem with fixed demands and dedicated storage. The scheduling model is described in a digraph form where nodes corresponded to possible task-unit-time allocations, and arcs corresponded to the possible sequences of tasks. The explicit description of the sequence in this form addressed one of the major weaknesses of the discrete-time model of Kondili et al. (1993). The formulation of Gooding et al. (1994) performed relatively well in problems with a strong sequencing component, but suffered from model complexity in that all possible sequences must be accounted for directly.

Pantelides et al. (1995) reported a STN-based approach to the scheduling of pipeless plants, where the material is conveyed between processing stations in movable vessels, and thus requiring the simultaneous scheduling of the movement and processing operations.

Blömer and Günther (1998) proposed a series of LP-based heuristics that can reduce solution times considerably, without compromising the quality of the solution obtained. Rodrigues et al. (2000) addressed the short-term planning/scheduling problems when the product demands are driven by customer orders. They proposed a multilevel decomposition procedure, containing at least two levels. At the planning level, demands were adjusted, a raw material delivery plan was defined and a capacity analysis was performed. Therefore, time windows for each operation were defined. At the scheduling level, an STN-based MIP model was developed. Grunow et al. (2002) show how the STN tasks could be aggregated into higher level processes for the purposes of longer term campaign planning.

Pantelides (1994) presented a critique of the STN and associated scheduling formulations, and he introduced the RTN process representation in order to overcome the drawbacks of the STN. He developed a discrete-time model based on the RTN which, due to the uniform treatment of resources, only required the description of three types of constraint (i.e., task allocation, batch size, and resource availability), and does not distinguish between identical units; a fact that resulted in more compact and less degenerate optimization models. He also demonstrated that the integrality gap could not be worse than the most efficient form of STN formulation.

At this point, it should be emphasized that while the discrete-time STN and RTN models are quite general and effective in monitoring the level of limited resources at the fixed times, their major weakness is the handling of long time horizons and relatively small processing and changeover times. Regarding the objective function, these models can easily handle profit maximization (cost minimization) for a fixed time horizon. Other objectives such as makespan minimization are more complex to implement since the time horizon and, in consequence, the number of time intervals required, are unknown a priori (Maravelias and Grossmann 2003). For these reasons, the more recent research has been focused mainly on developing scheduling models based on a continuous representation of time, where fewer grid points are required as they will be placed at the appropriate resource utilization discontinuities during problem solution.

**Continuous-Time Models**

A number of mathematical programming approaches have been developed for the scheduling of multiproduct batch plants. All are based (either explicitly or implicitly) on a continuous representation of time.

Pekny et al. (1988) addressed the scheduling problem in a multiproduct plant with no storage (i.e., ZW policy), and they show that the problem in question has the same structure as the asymmetric traveling salesman problem. They applied an exact parallel computation technique employing a tailor-made branch-and-bound procedure that used an assignment problem to provide problem relaxations. Then, Pekny et al. (1990) extended this work in order to account for product transition costs. Linear programming relaxations were used, and large-scale problems were solved to optimality with relatively modest computational effort. Finally, Gooding et al. (1994) further extended this work to cover the case of multiple units at each processing stage.

Birewar and Grossmann (1989) developed a MIP formulation for a similar type of plant, and they demonstrated that a straightforward LP model could be used to minimize the makespan: (i) through careful modeling of slack times and (ii) by exploiting the fact that relatively large numbers of batches of relatively few products will be produced. The result was a family of schedules, from which an individual schedule may be extracted. Birewar and Grossmann (1990) further extended this work for simultaneous long-term planning and scheduling, where the planning function took account of scheduling limitations.

Pinto and Grossmann (1995) proposed a MIP model for the minimization of earliness in a multiproduct plant with multiple units at each processing stage. Two types of individual time grids were used: one for units and one for orders. For each unit, a number of intervals of unknown duration were defined, which represented the possible sequence of tasks (one per interval). For each order, the time interval corresponded to a processing stage. These interval durations were also unknown, since processing times were unit-dependent. In order to ensure that, when a stage of an order was assigned to a unit, the starting times on both grids were equal, a set of mixed integer constraints was used. Precedence relations were employed for the material balances. Afterward, Pinto and Grossmann (1997) extended this model to take into account of the interactions between processing stages and shared resources (e.g., steam). They retained the individual grids, and account for the resource discontinuities through complex mixed integer constraints, which weakened the model and resulted in large computational times. For this reason, they developed a hybrid logic-based/MIP algorithm, where the disjunctions related to the relative timing of orders, in order to reduce the computational cost.

Moon et al. (1996) developed a MIP formulation for ZW multiproduct plants. The objective was to assign tasks to sequence positions so as to minimize the makespan, with nonzero transfer and setup times being included. Afterward, Kim et al. (1996) further extended this work for more general intermediate storage policies. They developed several formulations based on completion time relations.

Cerdá et al. (1997) and Karimi and McDonald (1997) addressed the case of single-stage processes with multiple units per processing stage. Cerdá et al. (1997) focused on changeovers and order fulfillment, while Karimi and McDonald (1997) focused on semicontinuous processes and total cost (i.e., transition, shortage and inventory) with the complication of minimum run lengths. A characteristic of both approaches is that discrete demands must be captured on the continuous-time grid. Méndez et al. (2000) developed a continuous-time precedence-based MIP model for a process with a single production stage with parallel units followed by a storage stage with multiple units, with restricted connectivity between the stages.

Afterward, Méndez et al. (2001) further extended this work to the multistage case with discrete shared resources. In common with other models, there were no explicit time slots in the model, and the key variables were allocations of activities to units and the relative orderings of activities.

Hui and Gupta (2001) presented a MIP formulation for the short-term scheduling of multiproduct batch plants with parallel no identical production units. They used bi index, instead of typical tri-index, discrete decision variables. As a result the number of discrete variables was decreased; however, the number of constraints was increased. Lee et al. (2002) developed a MIP model for scheduling problems in single-stage and continuous multiproduct processes on parallel lines with intermediate due dates and restrictions on minimum run lengths. Chen et al. (2002) proposed a continuous-time MIP model for the short-term scheduling of multiproduct single-stage batch plants with parallel lines involving constraints concerning release times and due dates of orders, as well as the sequence-dependent setup times and forbidden subsequences of production orders and the ready times of units. They also introduced some heuristic rules, and demonstrated that the rational employment of these heuristic rules could cut down the size of the model and had no effect on the optimality of the scheduling problem.

Chen et al. (2008) studied the medium-term planning problem of a single-stage single-unit continuous multiproduct polymer plant. They proposed a slot-based MIP model based on a hybrid discrete/continuous-time representation, where the production planning horizon was divided into several discrete weeks, and each week was formulated with a continuous-time representation. Liu et al. (2008) further improved the work of Chen et al. (2008) by presenting a MIP formulation based on the classic traveling salesman problem formulation. Their proposed model, without time slots, was computationally more effective. Recently, Liu et al. (2009) extended that work for medium-term planning of a single-stage multiproduct continuous plant to the case with parallel units. Erdirik-Dogan and Grossmann (2006) proposed a multiperiod slot-based MIP model for the simultaneous planning and scheduling of single-stage single-unit multiproduct continuous plants. A bi-level decomposition algorithm in which the original problem is decomposed into an upper level planning and a lower level scheduling problem was also developed in order to deal with complex problems. Erdirik-Dogan and Grossmann (2008) later extended their work to address parallel units. Sung and Maravelias (2008) presented a MIP formulation for the production planning of single-stage multiproduct processes. The problem was formulated as a multi-item capacitated lot-sizing problem in which: (i) multiple items can be produced in each planning period, (ii) sequence independent setups can carryover from previous periods, (iii) setups can crossover planning period boundaries, and (iv) setups can be longer than one period.

In the literature, some RTN-based continuous formulations have been reported. Specifically, Castro et al. (2001) developed a MIP formulation for the optimal scheduling of batch processes. Their formulation used a continuous-time representation and is based on the RTN representation. Castro and Grossmann (2006) presented a multiple-time grid RTN-based continuous-time MIP model for the short-term scheduling of single-stage multiproduct batch plants, which was based on the general formulation proposed by Castro et al. (2004). The most important difference was that a different time grid was used for each machine of the process instead of a single time grid for all events taking place. Their model can handle both release and due dates while the objective can be either the minimization of total cost or total earliness. Castro et al. (2006) developed two multiple-time-grid continuous-time MIP models for the scheduling of multiproduct multistage plants featuring equipment units with sequence-dependent changeovers. The performance of both formulations was compared to other MIP models and constraint program- ming models. The results show: (i) that multiple-time-grid models were better suited for single-stage problems or, when minimizing total earliness, (ii) that the constraint programming model was the best approach for makespan minimization, and (iii) that the continuous-time model with global precedence variables was the best overall performer.

A number of continuous-time formulations based both on the STN or the RTN representation and the definition of global time points have been developed. Mockus and Reklaitis (1999a, b) presented a general STN-based mathematical framework for describing scheduling problems arising in multipurpose batch and continuous chemical plants. The problem was formulated as a large nonlinear MIP model. A technique that exploits the characteristics of the problem in order to reduce the amount of required computation was also reported. Schilling and Pantelides (1996) presented a general MIP formulation for optimal scheduling of processes. Their continuous-time formulation was based on the RTN representation. In common with other continuous-time scheduling formulations, this exhibited a large integrality gap that rendered its solution using standard branch-and-bound algorithms highly problematic. Therefore, a branch-and-bound algorithm that branched on both discrete and continuous variables was proposed to address this complication. Zhang and Sargent (1996) extended the RTN concept in order to provide a unified mathematical formulation of the problem of determining the optimal operating conditions of a mixed production facility, comprising multipur- pose plant for both batch and continuous operations. Their formulation used a variable event time sequence common to all system events. This resulted into a large nonlinear MIP problem. However, for batch processes with fixed recipes, the problem was linear and can be solved by existing techniques.

Giannelos and Georgiadis (2002) developed an STN-based MIP formulation for scheduling multipurpose batch processes. A number of event points were prepos- tulated, which was the same for all tasks in the process. The ends of task execution defined event times, and they are generally different for different tasks of the process, giving rise to a no uniform time grid. The necessary time monotonic for single tasks was ensured by means of simple duration constraints. Suitable sequencing constraints, applicable to batch tasks involving the same state, were also introduced, so that state balances were properly posed in the context of the no uniform time grid. The expression of duration and sequencing constraints was greatly simplified by hiding all unit information within the task data. Their model was less computationally expensive from the previously reported models mainly due to smaller model size.

Maravelias and Grossmann (2003) developed an STN-based continuous-time MIP model for the scheduling of multipurpose batch plants. Their model addressed the general problem of batch scheduling, accounting for resource constraints, variable batch sizes and processing times, various storage policies (i.e., UIS, FIS, NIS, and ZW), batch mixing/splitting, and sequence-dependent changeover times. The key features of their model were: (i) a continuous-time representation was used, common for all units, (ii) assignment constraints were expressed using binary variables that were defined only for tasks, not for units, (iii) start times of tasks were eliminated, so that time-matching constraints were used only for the completion times of tasks, and (iv) a new class of valid inequalities that improved the LP relaxation was added to the MIP formulation. Maravelias (2005) proposed a mixed-time representation for STN-based scheduling models, where the time grid was fixed, but processing times were allowed to be variable and span an unknown number of time periods. The proposed representation was able to handle batch and continuous processes, and optimized holding, backlog, and utility costs. It also dealt with the release and due dates at no additional computational cost, and coped with variable processing times. It is also worth mentioning a recent study by Ferrer-Nadal et al. (2008) that incorporated the representation of transfer times, which had been ignored in STN- and RTN-based formulations thus not guaranteeing the generation of feasible solutions. By considering transfer times the generation of infeasible solutions, previously reported in the literature, was avoided.

Prasad and Maravelias (2008) developed a MIP formulation that involved three levels of discrete decisions, i.e., selection of batches, assignment of batches to units, and sequencing of batches in each unit. Continuous decision variables included sizing and timing of batches. They considered various objective functions: minimization of makespan, earliness, lateness, and production cost, as well as maximization of profit, an objective not addressed by previous multistage scheduling methods. In addition, in order to enhance the solution of their model, they proposed symmetry breaking constraints, developed a preprocessing algorithm for the generation of constraints that reduced the number of feasible solutions, and fixed sequencing variables based upon time window information. Sundaramoorthy and Maravelias (2008b) extended the work of Prasad and Maravelias (2008) to account for variable processing times. To account for batching decisions, they used additional batch selection and batch size variables and introduce demand satisfaction and unit-capacity constraints. Assignment constraints were active only for the subset of batches that were selected, and sequencing was carried out between batches that were assigned on the same processing unit. They also proposed an alternate formulation to handle sequence-dependent changeover costs. Finally, they presented methods that allowed to fix a subset of sequencing variables as well as they developed a set of tightening inequalities based on time windows, in order to enhance the computational performance of their model.

The works of Prasad and Maravelias (2008) and Sundaramoorthy and Maravelias (2008b) assumed unlimited storage. Méndez and Cerdá (2003b) and Wu and He (2004) considered storage constraints for scheduling in the more general multipurpose batch processes, however, did not consider batching decisions. Sundaramoorthy and Maravelias (2008a) proposed a precedence-based MIP formulation for the simultaneous batching and scheduling in multiproduct multistage processes with storage constraints, and they showed how their model could be modified to address all storage policies. They also discussed a class of tightening constraints, and they presented an extension for the modeling of changeover costs.

Gimenez et al. (2009a) presented a network-based MIP framework for the short-term scheduling of multipurpose batch processes. Their approach was based on five key concepts: (i) a new continuous-time representation is developed that does not require tasks to start (end) exactly at a time point; thus reducing the number of time points needed to represent a solution, (ii) processing units were modeled as being in different activity states to allow storage of input/output materials, (iii) time variables for "idle" and "storage" periods of a unit were introduced to enable the matching between tasks and time points without big-M constraints, (iv) material transfer variables were added to explicitly account for unit connectivity, and (iv) inventory variables for storage in processing units were incorporated to model nonsimultaneous and partial material transfers. Afterward, Gimenez et al. (2009b) extended this work to address aspects such as (i) preventive maintenance activities on unary resources (e.g., processing and storage units) that were planned ahead of time, (ii) resource-constrained changeover activities on processing and shared storage units, (iii) non instantaneous resource-constrained material transfer activities, (iv) intermediate deliveries of raw materials and shipments of finished products at predefined times, and (v) scenarios where part of the schedule was fixed because it had been programmed in the previous scheduling horizon.

Marchetti and Cerdá (2009a) presented a MIP continuous-time approach for the scheduling of single-stage multiproduct batch plants with parallel units and sequence-dependent changeovers. Their formulation was based on a unit-specific precedence-based representation. By explicitly including the equipment index in the domain of the sequencing variables, additional nontrivial tightening constraints producing better lower bounds on the optimal values of alternative objective functions (i.e., makespan or overall earliness) or key variables (i.e., task starting and completion times) were developed. Marchetti et al. (2010) proposed two precedence-based MIP continuous-time formulations (i.e., a rigorous and a cluster-based MIP) for the simultaneous lot-sizing and scheduling of single-stage multiproduct batch facilities. Both approaches can handle multiple customer orders per product at different due dates as well as variable processing times. The two proposed models differ in the way that sequencing decisions were taken. The rigorous approach dealt with the sequencing of individual batches processed in the same unit, while the approximate cluster-based method arranged groups of batches, each one featuring the same product, due date, and assigned unit. Since cluster members were often consecutively processed, each cluster can be treated and assigned to units as a single entity for sequencing purpose. It should be noted that the cluster-based model might result in suboptimal solutions.

**Multisite Production**

Much of the research effort to date has focused on the planning and scheduling of production for individual plants situated at a single geographical site and involving a set of batch, semicontinuous, or even continuous unit operations. As is well known, this is in itself a complex problem, optimal, or even feasible solutions to which are often notoriously difficult to obtain. However, it must also be recognized that production scheduling is only one aspect of the wider problem of process scheduling. For instance, the scheduling of plant maintenance operations, the coordinated planning of the production at a number of distinct geographical locations, and the management of distribution and SCs, all lead to important scheduling problems that interact strongly with production scheduling at individual production plants. It might be expected that large benefits would ensue from coordinated planning across sites, in terms of costs and market effectiveness. Most business processes dictate that a degree of autonomy is required at each manufacturing and distribution site, but pressures to coordinate responses to global demand while minimizing cost imply that simultaneous planning of production and distribution across plants and warehouses should be undertaken. This would result in the most efficient utilization of all resources. A target-setting approach, where central plans set achievable production targets without imposing operational details is compatible with operational details being determined at each site.

Wilkinson et al. (1996) showed how the RTN representation of Pantelides (1994) could be used to represent a variety of distribution options. The multisite planning problem can, therefore, be directly posed using the RTN representation and the discrete-time model of Pantelides (1994). Wilkinson et al. (1996) recognized that a potential problem with this approach is the very large model sizes that will ensue. A secondary issue is that the development of a central plan to a very fine level of detail is probably unnecessary. This led to the development of an aggregation procedure by Wilkinson et al. (1995). The aim was to capture production and distribution capacities accurately without considering detailed scheduling. The same authors applied this technique to a continent-wide industrial case study. This involved optimally planning the production and distribution of a system with 3 factories and 14 market warehouses and over a 100 products.

Karimi and McDonald (1997) described a similar problem for multiple facilities that effectively produced products on single-stage continuous lines for a number of geographically distributed customers. Their basic model was of multiperiod LP form, and took account of the available processing time on all lines, transportation costs and shortage costs.

Timpe and Kallrath (2000) and Kallrath (2002b) described a general MIP model based on a time-indexed formulation covering the relevant features required for the complete Supply Chain Management (SCM) of a multisite production network. The model combined aspects related to production, distribution and marketing and involves production sites and sales points. Besides standard features of lot-sizing problems (raw materials, production, inventories, demands) further aspects, e.g., different time scales attached to production and distribution, the use of periods with different lengths, the modeling of batch and campaign production need to be considered. While the actual application was taken from the chemical industry, the model provided a starting point for many applications in the chemical process industry, food or consumer goods industry.

Verderame and Floudas (2009) presented a multisite operational planning model that provided daily production and shipment profiles that represented a tight upper bound on the true capacity of the SC under investigation. The proposed scheme effectively modeled the production capacity of each production facility within the SC. The proposed planning model was to an industrial case and favorably compared to an existing planning model.

The SCM problem is particularly challenging because it not only encompasses the decisions of the planning/scheduling levels described about but also distribution logistics, market and price uncertainties as well as financial aspects. A substantial amount of work is appearing in this respect: detailed scheduling considerations in SC design (Puigjaner et al. 2009; Li and Ierapetritou 2010), embedded financial issues and environmental aspects (Laínez et al. 2007; Puigjaner and Guillén-Gosálbez 2008; Bojarski et al. 2009) and the linking of marketing and SC models (Laínez et al. 2010b).

**Resource-Constrained Production**
Manufacturing resources are generally grouped into two types: *renewable* and *nonrenewable* resources. A renewable resource is one that is recovered when the task to which it was allocated has concluded. Renewable resources can be discrete (e.g., tools, manpower) or continuous (e.g., heating, refrigeration, and electricity). In contrast, nonrenewable resources, like intermediates or raw materials, are consumed by tasks and every resource capacity allocated to them is no longer recovered at their completion. The literature in resource-constrained scheduling and planning problems in process industries is rather poor.

Pinto and Grossmann (1997) presented a MIP sequential approach based on a slot-based continuous-time representation that extended a former mathematical formulation for unconstrained multistage batch plants (Pinto and Grossmann 1995). As the number of binary variables and big-M constraints substantially increased, the general MIP resource-constrained model became almost computationally unsolvable. Consequently, the authors developed a problem solution methodology that combined a branch-and-bound MIP algorithm with disjunctive programming. Lamba and Karimi (2002) and Lim and Karimi (2003) presented lot-based representations to tackle semicontinuous scheduling problems of single-stage parallel production lines with resource constraints. Lamba and Karimi (2002) used identical slots across all processors while Lim and Karimi (2003) employed asynchronous slots. Since the underlying idea of an asynchronous slot is similar to the unit-specific time event, checkpoints for resource utilization were placed at the start of each slot and additional variables and constraints should be included to establish the slot relative positions.

Méndez and Cerdá (2002a) developed a precedence-based MIP continuous-time representation that independently handles unit allocation and task sequencing decisions through different sets of binary variables. Sequencing variables allowed ordering the tasks allocated either to the same equipment unit or to another discrete resource. In this way, an important saving in binary variables was achieved. Afterward, Méndez and Cerdá (2002b) reported a more general MIP formulation to deal with both continuous and discrete finite renewable resources. Each continuous resource was divided into a discrete number of sub-sources or pieces that were assigned to tasks through new allocation variables. Then, sequencing variables were still used to ordering tasks allocated to the same discrete or continuous resource item. The maximum number of pieces into which a continuous renewable can be divided was a model parameter, while each piece capacity was a nonnegative variable selected by the model. However, the proposed resource representation may sometimes exclude the problem optimum from the feasible space and, consequently, optimality was not guaranteed. It should be pointed out that the models of Méndez and Cerdá (2002a) and Méndez and Cerdá (2002b) could potentially lead to overestimation of utility levels.

Sundaramoorthy et al. (2009) proposed a discrete-time MIP model for the simultaneous batching and scheduling in multiproduct multistage processes under utility (e.g., cooling water, steam, and electricity) constraints. Since different tasks often share the limited utilities at the same time, they used a common time-grid approach. Further, the proposed method handles the batching decisions (i.e., the number and sizes of batches) seamlessly without the usage of explicit batch selection variables. Finally, they introduce a new class of inventory variables and constraints, in order to preserve batch identity in storage vessels.

Marchetti and Cerdá (2009b) presented a general precedence-based MIP framework to the short-term scheduling of multistage batch plants that accounted for sequence-dependent changeover times, intermediate due dates and limited availability of discrete and continuous renewable resources. Their formulation relied on a continuous-time formulation based on the general precedence notion that uses different sets of binary variables to handle allocation and sequencing decisions. To avoid resource overloading, additional constraints in terms of sequencing variables and a new set of 0–1 overlapping variables were presented. Finally, pre-ordering rules can be easily implemented in the MIP model.

## 2.5.2 Food Process Industry

A plethora of contributions addressing production scheduling and planning problems can be found in the OR and PSE communities literature. However, the use of optimization-based techniques for scheduling food process industries is still in its infancy. This can be mainly attributed to the complex production recipes, the large number of products to be produced under tight operating and quality constraints and the existence of mixed batch and semicontinuous production modes.

The literature in the field of single-site production scheduling and planning of food processing industries is rather poor. Entrup et al. (2005) presented three different MIP model formulations, which employed a combination of a discrete- and continuous-time representation, for scheduling and planning problems in the packing stage of stirred yogurt production. They accounted for shelf life issues and fermentation capacity limitations. However, product changeover times and production costs were ignored. The latter makes the proposed models more appropriate to cope with planning rather than scheduling problems, where products change-overs details are crucial. The data set used to demonstrate the practical applicability of their models consisted of 30 products based on 11 recipes that could be processed on four packing lines. They reported near-optimal solutions within the reasonable computational time for the case study solved.

Marinelli et al. (2007) addressed the planning problem of 17 products in 5 parallel packing machines, which share resources, in a packing line producing yogurt. The minimization of inventory, production and machines setup cost was their optimization goal. Sequence-dependent costs and times were not considered. They presented a discrete mathematical planning model that failed to obtain the optimal solution of the real application in an acceptable computation time. For this reason, they proposed a two-stage heuristic for obtaining near-optimal solutions for the problem under study.

Doganis and Sarimveis (2008) studied the scheduling problem at a yogurt packing line of a dairy company in Greece. Their objective was to optimally schedule two (or three) parallel conjoined (coupled) packing machines over a 5-day production horizon in order to meet the weekly demand for 25 different products. Each one of the identical machines could produce any of the 25 products. Products changeover times and costs were considered and total demand satisfaction was imposed. Simultaneous packing of multiple products was not allowed since the parallel machines shared the same feeding line. The latter restriction as well as the limited number of products considered greatly simplified the problem under question. The apparent reduction of changeover times was transformed into additional machine idle time. Finally, potential limitations of the fermentation stage were completely ignored.

The food production and distribution networks show a number of distinct features, such a sensitive quality of the products, production processes with both continuous and batch characteristics, the generation of by-products, and severe food safety and hygienic requirements (Grunow and van der Vorst 2010). Akkerman et al. (2010) recently presented an excellent review of quantitative operations management approaches to food distribution management, and relate this to challenges faced by the industry. A number of research challenges in strategic network design, tactical network planning, and operational transportation planning were highlighted with emphasis on food quality, food safety, and sustainability.

Brown et al. (2001) presented a large-scale linear program that modeled the production and distribution network of the Kellogg Company, a large producer of breakfast cereals and other foods. A salient aspect of the proposed model is that it was functioning on different time scales, using weeks or, months as time units.

Higgins et al. (2006) presented a model to schedule the shipment of sugar from production sites to ports from which ships were used to export sugar internationally. The main objective of this approach was to support rescheduling activities during the season to account for changing production rates. Eksioglu and Jin (2006) developed a general MIP approach for network planning of perishable products. Perishability was modeled by a maximum number of periods over which the product could be stored. A constraint was added in the formulation to make sure that product inventory in distribution centers was not used to cover the demand after having been stored beyond the specified maximum number of periods.

Ahumada and Villalobos (2009) critically reviewed the main contributions in the field of production and distribution planning for agri-foods based on agricultural crops. They focused on models that have been successfully implemented in problems of industrial interest. The models were classified according to relevant features, such as the optimization approaches used, the type of crops modeled and the scope of the plans, among many others.

Bilgen and Günther (2010) presented a so-called block planning approach that established cyclical production patterns based on the definition of setup families. Two transportation modes were considered for the delivery of final goods from the plants to distribution centers, full truckload, and less than truckload. The proposed MIP model minimized total production and transportation costs. A number of example problems illustrated the applicability of the proposed planning approach. Rong et al. (2009) described a MIP model that cleverly integrated food quality degradation in decision-making on production and distribution in food SCs.

Melo et al. (2009) presented an excellent and comprehensive review of the most recent literature contributions on facility location analysis within the context of SCM. They discussed the general relation between facility location models and strategic SC planning. A number of separate sections were dedicated to the relation between facility location and SCM as well as efficient solution methods and applications studies.

Manzini and Bindi (2009) presented an integrated framework for the design and optimization of a multi-echelon, multilevel production/distribution system. The framework relies on MIP techniques combined with cluster analysis, heuristic algorithms, and optimal transportation rules.

Recently Moula et al. (2010) presented a review of mathematical programming models for SC production and transport planning. The review critically identified current and future research in this field and proposed a taxonomy framework based on a number of elements such as SC structure, decision level, modeling approach, purpose, novelty, and applications.

### 2.5.3  Industrial Applications

The vast literature in the scheduling and planning area highlights the successful application of different optimization approaches to an extensive variety of challenging problems. As the economic advantages of implementing scheduling and planning tools became evident, BASF, DOW, and Du Pont began more intensive use of in-house developed tools for their planning and scheduling. Nowadays, more difficult and larger problems than those studied years ago can be solved sometimes even to optimality in a reasonable time by using more efficient integrated mathematical frameworks. This important achievement comes mainly from the remarkable advances in modeling techniques, algorithmic solutions and computational technologies that have been made in the last few years. Although a promising near future in the area can be predicted from this optimistic current situation, it is also well known that the actual gap between practice and theory is still evident. New academic developments are mostly tested on complex but relatively small problems whereas current real-world applications consist of hundreds of batches, dozens of pieces of equipment, and long scheduling periods, usually ranging from one to several weeks (Méndez et al. 2006).

Honkomp et al. (2000) gave a list of reasons why the practical implementation of scheduling tools based on optimization is fraught with difficulty. These include:

- The large amount of user defined input for testing purposes.
- The difficulty in capturing all the different types of operational constraints within a general framework, and the associated difficulty in defining an appropriate objective function.
- The large amounts of data required.
- Computational difficulties associated with the large problem sizes found in practice. Optimality gaps arising out of many shared resources.
- Intermediate storage and material stability constraints.
- Nonproductive activities (e.g., setup times and cleaning).
- Effective treatment of uncertainties in demands and equipment effectiveness.

However, there have been some success stories in the application of state-of-the-art scheduling and planning methods in the process industry. Espuña and Puigjaner (1989) reported an early success story in a complex problem in the textile sector. The detailed mathematical model built included specific ad hoc rules that permitted a fast simulation to obtain a feasible solution to start optimization. This strategy was applied to deal with the problem encountered in a large textile factory. Different fabrics and designs were produced using diverse machines, which could be shared by some products. The manufacturing of a total of 35,000 articles was considered. This situation resulted in a very complex problem of task assignment and optimization of production lines for the best use of the existing equipment to meet specified orders. Typical figures indicated that even qualified and experienced personnel found that drawing up the production plans for this kind of industrial facility was a burdensome job if it had to be done manually. The high-performance simulation module enabled the production manager to easily modify long-term and short-term production plans, and to evaluate objectively the consequences of such modifications (or decide to implement the changes suggested by the short-term production planning module). Hence, the production manager could cope with engineering decisions that require immediate attention. This feature was very useful when market pressures lead to the need for unexpected changes in a long-term production planning policy.

Schnelle (2000) applied MIP-based scheduling and design techniques for an agrochemical facility. The results indicated that sharing of equipment items between different products was a good idea, and the process reduced the number of alternatives to consider to a manageable number. Berning et al. (2002) described a large-scale planning/scheduling application that uses genetic algorithms for detailed scheduling at each site and a collaborative planning tool to coordinate plans across sites. The plants all operate batchwise, and may supply each other with intermediates, thus creating interdependencies in the plan. The scale of the problem was large, involving 600 different process recipes, and 1000 resources. Kallrath (2002a) presented the successful application of MIP methods for planning and scheduling in BASF. He described a software tool for simultaneous strategic and operational planning in a multisite production network. The total net profit of a global network was optimized, where key decisions included: (i) operating modes of equipment in each time period, (ii) production and supply of products, (iii) minor changes to the infrastructure (e.g., addition and removal of equipment from sites), and (iv) purchases and contracts of raw materials. A multiperiod model was formulated where equipment may undergo one mode change per period. The standard material balance equations were adjusted to account for the fact that transportation times are much shorter than the period durations. Counterintuitive but credible plans were obtained which resulted in cost savings of several millions of dollars. Keskinocak et al. (2002) described the application of a combined agent and optimization-based framework for the scheduling of paper products manufacturing. The frameworks solved the problems of order allocation, run formation, trimming and trim loss minimization, and load planning. The deployment of the system was claimed to save millions of dollars per year. Their approach used constructor and improver agents to generate candidate solutions that are evaluated against multiple criteria. Wang et al. (2010) and Harjunkoski and Grossmann (2001) addressed complex real-world scheduling problems in the polymer and the steel-making casting industry, respectively.

## 2.6  Trends and Challenges

A considerable amount of fruitful research work has already been carried out on scheduling and/or planning in process industries. Mathematical optimization can provide a quantitative basis for decisions and allow coping most successfully with complex problems, and it has proven itself as a useful technique to reduce costs and to support other objectives. Despite that, this technology has not yet found its way into many commercial software packages. For scheduling problems, there is not yet a commonly accepted state-of-the-art technology although some promising approaches have been developed, especially for job shop problems. Nevertheless, the majority of software packages are still based on pure heuristics (Kallrath 2002b).

A number of issues can provide interesting future research challenges in the research field of scheduling and/or planning in process industry. Based on the literature review, it is foreseen the need to devote further research efforts in order to meet the following trend and challenges:

- The recent research is all about solution efficiency and techniques to render tractable even larger problems. There remains work to be done on both model enhancements and improvements in solution algorithms if industrially relevant problems are to be tackled routinely, and software based on these are to be used on a regular basis by practitioners in the field.
- Much of the more recent research has focused on continuous-time formulations, but little technology has been developed based on these. The main challenge here is in continual improvement in problem formulation and preprocessing to improve relaxation characteristics, and tailored solution procedures for problems with relatively large integrality gaps.
- The mathematical models developed should be implemented into industrial or industrial-based studies, in order to demonstrate to industrial practitioners the potential benefits for adopting mathematical programming methods in managing scheduling and/or planning problem in industrial environments.
- The multisite problem has received relatively little attention, and is likely to be a candidate for significant research in the near future. A major challenge is to develop planning approaches that are consistent with detailed production scheduling at each site and distribution scheduling across sites. An obvious obstacle is the problem size, therefore appropriate modeling frameworks should be devised in order to tackle rigorously and efficiently these highly complicated optimization problems.
- There are some process industries that have received little attention; regarding scheduling and/or planning research. One of the most emerging and challenging industries of this type, is the food process industry. Scheduling and planning approaches in the food process industry is rather poor, despite the fact that there are many optimization challenges. There is a need for optimization frameworks able to cope with scheduling/planning problem under the complex semicontinuous process mode of these industries.
- Another focus of modeling, which is possible now due to increased computer power available, is the opportunity to solve design and operational planning problems, or strategic and operational planning problems simultaneously in one model.
- Another challenge relates to the seamless integration of the activities at different decision levels; this is of a much broader and more interdisciplinary nature. The financial aspects will require more rigorous treatment, as scheduling and planning become integrated.
- Another major issue is the handling of uncertainty (e.g., in terms of processing times, and availability of equipment, modification of orders). A major challenge here is how to best formulate a stochastic optimization model that is meaningful and whose results are easy to interpret and implement.

# References

Ahumada O, Villalobos JR (2009) Application of planning models in the agri-food supply chain: a review. Eur J Oper Res 195:1–20

Akkerman R, Farahani P, Grunow M (2010) Quality, safety and sustainability in food distribution: a review of quantitative operations management approaches and challenges. OR Spectrum 32:863–904

Aytug HM, Lawley M et al (2005) Executing production schedules in the face of uncertainties: a review and some future directions. Eur J Oper Res 161(1):86–110

Baptiste P, Le Pape C, Nuijten W (2001) Constraint-based scheduling. Applying constraint programming to scheduling problems. Kluwer Academic Publishers, Boston

Bassett MH, Pekny JF, Reklaitis GV (1997) Using detailed scheduling to obtain realistic operating policies for a batch processing facility. Ind Eng Chem Res 36:1717–1726

Berning G, Brandenburg MK et al (2002) An integrated system for supply chain optimisation in the chemical process industry. OR Spectrum 24(4):371–401

Bilgen B, Günther HO (2010) Integrated production and distribution planning in the fast moving consumer goods industry: a block planning application. OR Spectrum 32:927–955

Birewar D, Grossmann IE (1989) Efficient optimization algorithms for zero-wait scheduling of multiproduct batch plants. Ind Eng Chem Res 28(9):1333–1345

Birewar D, Grossmann IE (1990) Simultaneous planning and scheduling of multiproduct batch plants. Ind Eng Chem Res 29:570–580

Blackstone JH, Phillips DT, Hogg GL (1982) A state-of-the-art survey of dispatching rules for manufacturing job shop operations. Int J Prod Res 20:27–45

Blömer F, Günther HO (1998) Scheduling of a multi-product batch process in the chemical industry. Comput Ind 36(3):245–259

Blömer F, Günther HO (2000) LP-based heuristics for scheduling chemical batch processes. Int J Prod Res 38:1029–1051

Bojarski AD, Laínez JM, Espuña A, Puigjaner L (2009) Incorporating environmental impacts and regulations in a holistic supply chains modeling: an LCA approach. Comput Chem Eng 33 (10):1747–1759

Bonfill A (2006) Proactive management of uncertainty to improve scheduling robustness in process industries. Ph.D. thesis. Universitat Politecnica de Catalunya

Brown GJ, Keegan B et al (2001) The Kellogg company optimizes production, inventory, and distribution. Interfaces 31:1–15

Burkard RE, Hujter M et al (1998) A process scheduling problem arising from chemical production planning. Opt Methods Softw 10:175–196

Castro PM, Grossmann IE (2006) An efficient MILP model for the short-term scheduling of single stage batch plants. Comput Chem Eng 30(6–7):1003–1018

Castro P, Barbosa-Póvoa APFD, Matos HA (2001) An improved RTN continuous-time formulation for the short-term scheduling of multipurpose batch plants. Ind Eng Chem Res 40(9):2059–2068

Castro PM, Barbosa-Póvoa APFD et al (2004) Simple continuous-time formulation for short-term scheduling of batch and continuous processes. Ind Eng Chem Res 43(1):105–118

Castro PM, Grossmann IE, Novais AQ (2006) Two new continuous-time models for the scheduling of multistage batch plants with sequence dependent changeovers. Ind Eng Chem Res 45(18):6210–6226

Castro PM, Harjunkoski I, Grossmann IE (2009) Optimal short-term scheduling of large-scale multistage batch plants. Ind Eng Chem Res 48:11002–11016

Cavin L, Fischer U, Glover F, Hungerbhüler K (2004) Multi-objective process design in multipurpose batch plants using a tabu search optimization algorithm. Comput Chem Eng 28:459–478

Cerdá J, Henning GP, Grossmann IE (1997) A mixed-integer linear programming model for short-term scheduling of single-stage multiproduct batch plants with parallel lines. Ind Eng Chem Res 36(5):1695–1707

Chen CL, Liu CL et al (2002) Optimal short-term scheduling of multiproduct single-stage batch plants with parallel lines. Ind Eng Chem Res 41(5):1249–1260

Chen P, Papageorgiou LG, Pinto JM (2008) Medium-term planning of single-stage single unit multiproduct plants using a hybrid discrete/continuous-time MILP model. Ind Chem Eng Res 47:1925–1934

Dannenbring DG (1977) Evaluation of flow shop sequencing heuristics. Manage Sci 23:1174–1182

Doganis P, Sarimveis H (2008) Optimal production scheduling for the dairy industry. Ann Oper Res 159:315–331

Eksioglu SD, Jin M (2006) Cross-facility production and transportation planning problem with perishable inventory. Computational science and its applications—ICCSA 2006. Springer, Berlin, pp 708–717

Elkamel A (1993) Scheduling of process operations using mathematical programming techniques. Ph.D. thesis. Purdue University

Entrup ML, Günther HO et al (2005) Mixed-integer linear programming approaches to shelf-life-integrated planning and scheduling in yoghurt production. Int J Prod Res 43:5071–5100

Erdirik-Dogan M, Grossmann IE (2006) A decomposition method for the simultaneous planning and scheduling of single-stage continuous multiproduct plants. Ind Eng Chem Res 45:299–315

Erdirik-Dogan M, Grossmann IE (2008) Simultaneous planning and scheduling of single-stage multi-product continuous plants with parallel lines. Comput Chem Eng 32:2664–2683

Espuña A, Puigjaner L (1989) Solving the production planning problem for parallel multiproduct plants. Inst Chem Eng Symp Ser 114:15–25

Ferrer-Nadal S, Capón-García E, Méndez CA, Puigjaner L (2008) Material transfer operations in batch scheduling. A critical modeling issue. Ind Eng Chem Res 47(20):7721–7732

Franca PM, Gendreau M, Laporte G, Müller FM (1996) A tabu search heuristic for the multiprocessor scheduling problem with sequence dependent setup times. Int J Prod Econ 43:78–89

Giannelos NF, Georgiadis MC (2002) A simple new continuous-time formulation for shortterm scheduling of multipurpose batch processes. Ind Eng Chem Res 41(9):2178–2184

Gimenez DM, Henning GP, Maravelias CT (2009a) A novel network-based continuous-time representation for process scheduling: part I. Main concepts and mathematical formulation. Comput Chem Eng 33(9):1511–1528

Gimenez DM, Henning GP, Maravelias CT (2009b) A novel network-based continuous-time representation for process scheduling: part II. General framework. Comput Chem Eng 33 (10):1644–1660

Glover F (1990) Tabu search: a tutorial. Interfaces 20:74

Gooding WB (1994) Specially structured formulations and solution methods for optimisation problems important to process scheduling. Ph.D. thesis. Purdue University

Gooding WB, Pekny JF, McCroskey PS (1994) Enumerative approaches to parallel flowshop scheduling via problem transformation. Comput Chem Eng 18:909–927

Graells M, Espuña A, Puigjaner L (1996) Sequencing intermediate products: a practical solution for multipurpose production scheduling. Comput Chem Eng S20:S1137–S1142

Graells M, Cantón J, Peschaud B, Puigjaner L (1998) General approach and tool for the scheduling of complex production systems. Comput Chem Eng S22:S395–S402

Graves SC (1982) Using Lagrangean techniques to solve hierarchical production planning problems. Manage Sci 28:260–275

Grunow M, van der Vorst J (2010) Food production and supply chain management. OR Spectrum 32:861–862

Grunow M, Günther HO, Lehmann M (2002) Campaign planning for multi-stage batch processes in the chemical industry. OR Spectrum 24:281–314

Gupta A, Maranas CD (1999) A hierarchical Lagrangean relaxation procedure for solving midterm planning problems. Ind Eng Chem Res 38:1937–1947

Harjunkoski I, Grossmann IE (2001) A decomposition approach for the scheduling of a steel plant production. Comput Chem Eng 25(11):1647–1660

Henning GP, Cerdá J (2000) Knowledge-based predictive and reactive scheduling in industrial environments. Comput Chem Eng 24(9):2315–2338

Higgins A, Beashel G, Harrison A (2006) Scheduling of brand production and shipping within a sugar supply chain. J Oper Res Soc 57:490–498

Honkomp SJ, Lombardo S, Rosen O, Pekny JF (2000) The curse of reality—why process scheduling optimization problems are difficult in practice. Comput Chem Eng 24(2–7):323–328

Hui CW, Gupta A (2001) A bi-index continuous-time mixed-integer linear programming model for single-stage batch scheduling with parallel units. Ind Eng Chem Res 40(25):5960–5967

Ierapetritou MG, Floudas CA (1998) Effective continuous-time formulation for shortterm scheduling. 1. Multipurpose batch processes. Ind Eng Chem Res 37(11):4341–4359

Kallrath J (2002a) Combined strategic and operational planning—an MILP success story in chemical industry. OR Spectrum 24(3):315–341

Kallrath J (2002b) Planning and scheduling in the process industry. OR Spectrum 24:219–250

Karimi IA, McDonald CM (1997) Planning and scheduling of parallel semicontinuous processes. 2. Short-term scheduling. Ind Eng Chem Res 36(7):2701–2714

Keskinocak P, Wu F et al (2002) Scheduling solutions for the paper industry. Oper Res 50(2):249–259

Kim M, Jung JH, Lee IB (1996) Optimal scheduling of multiproduct batch processes for various intermediate storage policies. Ind Eng Chem Res 35(11):4048–4066

Kirkpatrick S, Gelatt CD, Vechi MP (1983) Optimization by simulated annealing. Science 220:671–680

Kondili E, Pantelides CC, Sargent RWH (1988) A general algorithm for scheduling of batch operations. In: Proceedings of third international symposium on process systems engineering. Sydney, Australia, pp 62–75

Kondili E, Pantelides CC, Sargent RWH (1993) A general algorithm for short term scheduling of batch operations. Comput Chem Eng 17:211–227

Ku H, Karimi IA (1991) An evaluation of simulated annealing for batch process scheduling. Ind Chem Eng Res 30:163–169

Kudva G, Elkamel A et al (1994) Heuristic algorithm for scheduling batch and semicontinuous plants with production deadlines, intermediate storage limitations and equipment changeover costs. Comput Chem Eng 18:859–875

Kuriyan K, Reklaitis GV (1989) Scheduling network flowshops so as to minimise makespan. Comput Chem Eng 13:187–200

Laborie P (2003) Algorithms for propagating resource constraints in AI planning and scheduling: existing approaches and new results. Artif Intell 143(2):151–188

Laínez JM, Guillén-Gonsálbez G et al (2007) Enhancing corporate value in the optimal design of chemical supply chains. Ind Eng Chem Res 46(23):7739–7757

Lainez JM, Hegyháti et al (2010a) Using S-graph to address uncertainty in batch plants. Clean Technol Environ Policy 12(2):105–115

Laínez JM, Reklaitis GV, Puigjaner L (2010b) Linking marketing and supply chain models for improved business strategic decision support. Comput Chem Eng 34(12):2107–2117

Lamba N, Karimi IA (2002) Scheduling parallel production lines with resource constraints. 1. Model formulation. Ind Eng Chem Res 41(4):779–789

Lázaro M, Puigjaner L (1988) Solution of integer optimization problems subjected to non-linear restrictions: an improved algorithm. Comput Chem Eng 12(5):443–448

Le Pape C (1998) Implementation of resource constraints in ILOG schedule: a library for the development of constrained-based scheduling systems. Intell Syst Eng 3(2):55–66

Lee KH, Heo S-K, Lee H-K, Lee L-B (2002) Scheduling of single-stage and continuous processes on parallel lines with intermediate due dates. Ind Eng Chem Res 41(1):58–66

Li Z, Ierapetritou MG (2008) Process scheduling under uncertainty: review and challenges. Comput Chem Eng 32(4–5):715–727

Li Z, Ierapetritou MG (2010) Production planning and scheduling integration through augmented Lagrangian optimization. Comput Chem Eng 34(6):996–1006

Lim M, Karimi IA (2003) Resource-constrained scheduling of parallel production lines using asynchronous slots. Ind Eng Chem Res 42(26):6832–6842

Liu S, Pinto JM, Papageorgiou LG (2008) A TSP-based MILP model for medium-term plan- ning of single-stage continuous multiproduct plants. Ind Chem Eng Res 47:7733–7743

Liu S, Pinto JM, Papageorgiou LG (2009) MILP-based approaches for medium-term planning of single-stage continuous multiproduct plants with parallel units. Comput Manage Sci 7:407–435

Majozi T, Friedler F (2006) Maximization of throughput in a multipurpose batch plant under a fixed time horizon: S-graph approach. Ind Eng Chem Res 45(20):6713–6720

Manzini R, Bindi F (2009) Strategic design and operational management optimization of a multi stage physical distribution system. Transp Res Part E Log Transp Rev 45(6):915–936

Maravelias CT (2005) Mixed-time representation for state-task network models. Ind Eng Chem Res 44(24):9129–9145

Maravelias CT, Grossmann IE (2003) New general continuous-time state-task network formulation for short-term scheduling of multipurpose batch plants. Ind Eng Chem Res 42(13):3056–3074

Marchetti PA, Cerdá J (2009a) A continuous-time tightened formulation for single-stage batch scheduling with sequence-dependent changeovers. Ind Eng Chem Res 48(1):483–498

Marchetti PA, Cerdá J (2009b) A general resource-constrained scheduling framework for multistage batch facilities with sequence-dependent changeovers. Comput Chem Eng 33(4):871–886

Marchetti PA, Méndez CA, Cerdá J (2010) Mixed-integer linear programming monolithic formulations for lot-sizing and scheduling of single-stage batch facilities. Ind Eng Chem Res 49:6482–6498

Marinelli F, Nenni ME, Sforza A (2007) Capacitated lot sizing and scheduling with parallel machines and shared buffers: a case study in a packaging company. Ann Oper Res 150:177–192

Melo MT, Nickel S, Saldanha-da-Gama F (2009) Facility location and supply chain management —a review. Eur J Oper Res 196:401–412

Méndez CA, Cerdá J (2002a) An MILP framework for short-term scheduling of single-stage batch plants with limited discrete resources. Comput Aided Chem Eng 10:721–726

Méndez CA, Cerdá J (2002b) Short-term scheduling of multistage batch processes subject to limited finite resources. Comput Aided Chem Eng 15B:984–989

Méndez CA, Cerdá J (2003a) Dynamic scheduling in multiproduct batch plants. Comput Chem Eng 27:1247–1259

Méndez CA, Cerdá J (2003b) An MILP continuous-time framework for short-term scheduling of multipurpose batch processes under different operation strategies. Opt Eng 4(1–2):7–22

Méndez CA, Henning GP, Cerdá J (2000) Optimal scheduling of batch plants satisfying multiple product orders with different due-dates. Comput Chem Eng 24(9–10):2223–2245

Méndez CA, Henning GP, Cerdá J (2001) An Milp continuous-time approach to short-term scheduling of resource constrained multistage flowshop batch facilities. Comput Chem Eng 25(4–6):701–711

Méndez CA, Cerdá J et al (2006) Review: state-of-theart of optimization methods for short-term scheduling of batch processes. Comput Chem Eng 30(6–7):913–946

Mockus L, Reklaitis GV (1999a) Continuous time representation approach to batch and continuous process scheduling. 1. MINLP formulation. Ind Eng Chem Res 38(1):197–203

Mockus L, Reklaitis GV (1999b) Continuous time representation approach to batch and continuous process scheduling. 2. Computational issues. Ind Eng Chem Res 38(1):204–210

Moon S, Park S, Lee WK (1996) New MILP models for scheduling of multiproduct batch plants under zero-wait policy. Ind Eng Chem Res 35:3458–3469

Moula J, Peidro D et al (2010) Mathematical programming models for supply chain production and transport planning. Eur J Oper Res 204:377–390

Mulvey JM, Vanderbei RJ, Zenios SA (1995) Robust optimization of large-scale systems. Oper Res 43(2):264–281

Murakami Y, Uchiyama H et al (1997) Application of repetitive SA method to scheduling problems of chemical processes. Comput Chem Eng S21:S1087–S1092

Pacciarelli D (2002) The alternative graph formulation for solving complex factory scheduling problems. Int J Prod Res 40:3641–3653

Pantelides CC (1994) Unified frameworks for optimal process planning and scheduling. In: Proceedings of second conference on foundations of computer aided process operations. CACHE, pp 253–274

Pantelides CC, Realff MJ, Shah N (1995) Short-term scheduling of pipeless batch plants. Chem Eng Res Des 73(A4):431–444

Panwalkar SS, Iskander WA (1977) Survey of scheduling rules. Oper Res 25:45–61

Papageorgiou LG, Pantelides CC (1996a) Optimal campaign planning/scheduling of multipurpose batch/semicontinuous plants. 1. Mathematical formulation. Ind Chem Eng Res 35(2):488–509

Papageorgiou LG, Pantelides CC (1996b) Optimal campaign planning/scheduling of multipurpose batch/semicontinuous plants. 2. A mathematical decomposition approach. Ind Chem Eng Res 35(2):510–529

Pekny JF, Miller DL, McCrae GJ (1988) Application of a parallel travelling salesman problem to no-wait flowshop scheduling. In: AIChE annual meeting 1988. Washington, DC, USA

Pekny JF, Miller DL, McCrae GJ (1990) An exact parallel algorithm for scheduling when production costs depend on consecutive system states. Comput Chem Eng 14:1009–1023

Pinedo M (1995) Scheduling, theory, algorithms and systems. Prentice-Hall, New York

Pinto JM, Grossmann IE (1995) A continuous time mixed integer linear programming model for short-term scheduling of multistage batch plants. Ind Eng Chem Res 34(9):3037–3051

Pinto JM, Grossmann IE (1997) A logic-based approach to scheduling problems with resource constraints. Comput Chem Eng 21(8):801–818

Pistikopoulos EN, Dua V et al (2002) On-line optimization via off-line parametric optimization tools. Comput Chem Eng 26(2):175–185

Prasad P, Maravelias CT (2008) Batch selection, assignment and sequencing in multi-stage multi-product processes. Comput Chem Eng 32(6):1106–1119

Puigjaner L (1999) Handling the increasing complexity of detailed batch process simulation and optimisation. Comput Chem Eng 23(S1):S929–S943

Puigjaner L, Guillén-Gosálbez G (2008) Towards an integrated framework for supply chain management in the batch chemical process industry. Comput Chem Eng 32(4–5):650–670

Puigjaner L, Laínez JM, Álvarez CR (2009) Tracking the dynamics of the supply chain for enhanced production sustainability. Ind Eng Chem Res 48(21):9556–9570

Raaymakers WHM, Hoogeveen JA (2000) Scheduling multipurpose batch process industries with no-wait restrictions by simulated annealing. Eur J Oper Res 126:131–151

Reklaitis GV (1982) Reviewing of scheduling of process operations. In: Chemical engineering progress symposium series: selected topics in computer aided process design and analysis. AIChE, New York, USA

Rippin DWT (1993) Batch process systems engineering: a retrospective and prospective review. Comput Chem Eng S17:S1–S13

Rodrigues MTM, Latre LG, Rodrigues LCA (2000) Short-term planning and scheduling in multipurpose batch chemical plants: a multi-level approach. Comput Chem Eng 24(9–10):2247–2258

Rong A, Akkerman R, Grunow M (2009) An optimization approach for managing fresh food quality throughout the supply chain. Int J Prod Econ. https://doi.org/10.1016/j.ijpe.2009.11.026

Röslof J, Harjunkoski I et al (2001) An MILP-based reordering algorithm for complex industrial scheduling and rescheduling. Comput Chem Eng 25:821–828

Ruiz R, Maroto C (2006) A genetic algorithm for hybrid flowshops with sequence dependent setup times and machine eligibility. Eur J Oper Res 169:781–800

Ruiz R, Stutzle T (2008) An iterated greedy heuristic for the sequence dependent setup times flowshop problem with makespan and weighted tardiness objectives. Eur J Oper Res 187:1143–1159

Sahinidis NV (2004) Optimization under uncertainty: state-of-the-art and opportunities. Comput Chem Eng 28(6–7):971–983

Sahinidis NV, Grossmann IE (1991) MINLP model for cyclic multiproduct scheduling on continuous parallel lines. Comput Chem Eng 15:85–103

Sanmartí E, Friedler F, Puigjaner L (1998) Combinatorial technique for short term scheduling of multipurpose batch plants based on schedule-graph representation. Comput Chem Eng 22(S1): S847–S850

Sanmartí E, Holczinger T, Friedler F, Puigjaner L (2002) Combinatorial framework for effective scheduling of multipurpose batch plants. AIChE J 48(11):2557–2570

Sauer J, Bruns R (1997) Knowledge-based scheduling in industry and medicine. IEEE Expert 12 (1997):24–31

Schilling G, Pantelides CC (1996) A simple continuous-time process scheduling formulation and a novel solution algorithm. Comput Chem Eng 20:S1221–S1226

Schnelle KD (2000) Preliminary design and scheduling of a batch agrochemical plant. Comput Chem Eng 24(2–7):1535–1541

Shah N, Pantelides CC, Sargent RWH (1993a) A general algorithm for short-term scheduling of batch operations—II. Computational issues. Comput Chem Eng 17:229–244

Shah N, Pantelides CC, Sargent RWH (1993b) Optimal periodic scheduling of multipurpose batch plants. Ann Oper Res 42:193–228

Shobrys DE (2001) The history of APS. Report of the supply chain consultants. Available at http://www.thesupplychain.com/

Sundaramoorthy A, Karimi IA (2005) A simpler better slot-based continuous-time formulation for short-term scheduling in multipurpose batch plants. Chem Eng Sci 60(10):2679–2702

Sundaramoorthy A, Maravelias CT (2008a) Modeling of storage in batching and scheduling of multistage processes. Ind Eng Chem Res 47(17):6648–6660

Sundaramoorthy A, Maravelias CT (2008b) Simultaneous batching and scheduling in multistage multiproduct processes. Ind Eng Chem Res 47(5):1546–1555

Sundaramoorthy A, Maravelias CT, Prasad P (2009) Scheduling of multistage batch processes under utility constraints. Ind Eng Chem Res 48(13):6050–6058

Sung C, Maravelias CT (2008) A mixed-integer programming formulation for the general capacitated lot-sizing problem. Comput Chem Eng 32(1–2):244–259

Takahashi K, Fujii H (1991) New concept for batchwise specialty chemicals production plant. Instrum Control Eng 1(2):19–22

Timpe CH, Kallrath J (2000) Optimal planning in large multi-site production networks. Eur J Oper Res 126:422–435

Van Hentenryck P (1989) Constraint satisfaction in logic programming. MIT Press, Cambridge

Van Hentenryck P (2002) Constraint and integer programming in OPL. INFORMS J Comput 14 (4):345–372

Venditti L, Pacciarelli D, Meloni C (2010) A tabu search algorithm for scheduling pharmaceutical packaging operations. Eur J Oper Res 202:538–546

Verderame PM, Floudas CA (2009) Operational planning framework for multisite production and distribution networks. Comput Chem Eng 33:1036–1050

Wang K, Löhl T, Stobbe M, Engell S (2010) A genetic algorithm for online-scheduling of a multiproduct polymer batch plant. Comput Chem Eng 24(2–7):393–400

Wilkinson SJ, Shah N, Pantelides CC (1995) Aggregate modelling of multipurpose plant operation. Comput Chem Eng S19:S583–S588

Wilkinson SJ, Cortier A et al (1996) Integrated production and distribution scheduling on a european-wide basis. Comput Chem Eng S20:S1275–S1280

Wu J, He X (2004) A new model for scheduling of batch process with mixed intermediate storage policies. J Chin Inst Chem Eng 35:381–387

Xia Q, Macchietto S (1994) Routing, scheduling and product mix optimization by minimax algebra. Chem Eng Res Des 72:408–414

Yee KL, Shah N (1997) Scheduling of fast-moving consumer goods plants. J Oper Res Soc 48:1201–1214

Yee KL, Shah N (1998) Improving the efficiency of discrete-time scheduling formulations. Comput Chem Eng S22:S403–S410

Zhang X, Sargent RWH (1996) The optimal operation of mixed production facilities—a general formulation and some approaches for the solution. Comput Chem Eng 20:897–904

Zweben M, Fox MS (1994) Intelligent scheduling. Morgan Kaufmann, San Francisco

# Chapter 3
# Methods and Tools

## 3.1 Mathematical Programming

In Chap. 2, a state-of-the-art review is carried out, which finally allows identifying some production scheduling and planning trends and challenges. Now, in this chapter, the methods and tools used throughout this book are briefly outlined.

Mathematical programming is the use of mathematical models, particularly optimizing models, to assist in taking decisions. The term "Programming" antedates computers and means "preparing a schedule of activities". It is still used, for instance, in oil refineries, where the refinery programmers prepare detailed schedules of how the various process units will be operated and the products blended. Mathematical programming is, therefore, the use of mathematics to assist in these activities. Mathematical Programming is one of a number of Operational Research (OR) techniques. Its particular characteristic is that the best solution to a model is found automatically by optimization software. A mathematical programming model answers the question "What's best?" rather than "What happened?" (Statistics), "What if?" (Simulation), "What will happen?" (Forecasting), or "What would an expert do and why?" (Expert systems).

Being so ambitious does have its disadvantages. Mathematical programming is more restrictive in what it can represent than other techniques. **Nor should it be imagined that it really does find the best solution to the real-world problem. It finds the best solution to the problem as modeled.** If the model has been built well, this solution should translate back into the real world as a good solution to the real-world problem. If it does not, analysis of why it is no good leads to greater understanding of the real-world problem.

Whatever the real-world problem is, it is usually possible to formulate the optimization problem in a generic form. All optimization problems with explicit objectives can, in general, be expressed as nonlinearly constrained optimization problems in the following generic form:

© Springer Nature Switzerland AG 2019
G. M. Kopanos and L. Puigjaner, *Solving Large-Scale Production Scheduling and Planning in the Process Industries*,
https://doi.org/10.1007/978-3-030-01183-3_3

$$\text{Maximize/minimize} f(x)$$
$$x \in \mathbb{R}^\times$$
$$\text{subject to} \tag{3.1}$$
$$\phi_m(x) \quad = 0 \quad (m = 1, \ldots, M)$$
$$\psi_k(x) \quad \leq 0 \quad (k = 1, \ldots, K)$$

$$\text{where } x = (x_1, x_2, \ldots, x_n)^T \in \mathbb{R}^n$$

where $f(x)$, $\phi_m(x)$, and $\psi_k(x)$ are scalar functions of the real column vector $x$. The function $f(x)$ is called *objective function*, and is a quantitative measure of the performance of the system in question. The components $x_i$ of vector $x$ are called *decision variables*, or simply variables, and they can be either continuous, discrete or a mixed of these two. The variables are the unknowns whose values are to be determined such that the objective functions is optimized. Additionally, $\phi_m(x)$ are constraints in terms of M equalities, and $\psi_k(x)$ are constraints written as K inequalities. Therefore, there are M + K constraints in total. Constraints represent any restrictions that the decision variables must satisfy.

The procedure of identifying the aforementioned components is known as modeling. Depending on the properties of the functions $f$, $\phi$, $\psi$, and the vector $x$, the mathematical program (3.1) is called:

- *Linear*: If $x$ is continuous and the functions $f$, $\phi$, and $\psi$, are all linear.
- *Nonlinear*: If $x$ is continuous and at least one of the functions $f$, $\phi$, and $\psi$, is nonlinear.
- *Mixed integer linear*: If $x$ requires at least some of the variables $x_i$ to take integer (or binary) values only; and the functions $f$, $\phi$, and $\psi$, are linear.
- *Mixed integer nonlinear*: If $x$ requires at least some of the variables $xi$ to take integer (or binary) values only; and at least one of the functions $f$, $\phi$, and $\psi$, is nonlinear.

### 3.1.1   Optimality Criteria

A point $x$ which satisfies all the constraints is called a feasible point and therefore is a *feasible solution* to the problem. The set of all feasible points is called the *feasible region*. A point $x_*$ is called a strong local maximum of the optimization problem if $f(x_*)$ is defined in a $\delta$-neighborhood $N(x_*, \delta)$ and satisfies $f(x_*) > f(u)$ for $\forall u \in N(x_*, \delta)$ where $\delta > 0$ and $u \neq x_*$. If $x_*$ is not a strong local maximum, the inclusion of equality in the condition $f(x_*) \geq f(u)$ for $\forall u \in N(x_*, \delta)$ defines the point $x_*$ as a weak local maximum (see Fig. 3.1). The local minima can be defined in the similar

**Fig. 3.1** Illustrate example for strong and weak maxima and minima

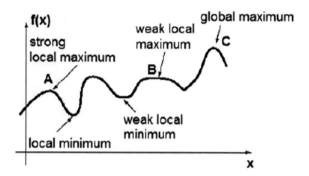

manner when > and ≥ are replaced by < and ≤, respectively. Figure 3.1 illustrates several local maxima and minima. Point A is a strong local maximum, and point B is a weak local maximum since there exist many different values of $x$ that will lead to the same value of $f(x_*)$. Finally, point C is a global maximum.

### 3.1.2 Convexity

Let $\zeta$ be a set in a real or complex vector space. Set $\zeta$ is *convex* if, for every pair of points $x$ and $y$ belonging within the set, every point on the straight line segment that connects them is also within the set $\zeta$, as illustrated in Fig. 3.2. This definition is can be mathematically expressed as:

$$\zeta \text{ is convex} \Leftrightarrow \forall (x, y) \in \zeta \land \theta \in [1, 0] : ((1 - \theta)x + \theta y) \in 2\zeta$$

A function $f(x)$ is convex if its epigraph (i.e., the set of points lying on or above its graph) is a convex set, as shown in Fig. 3.3. Convexity plays a significant role in mathematical programming due to the following theorem:

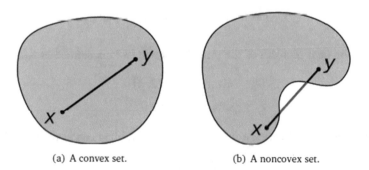

(a) A convex set.  (b) A noncovex set.

**Fig. 3.2** Graphical representation for convexity

**Fig. 3.3** Graphical
representation for a convex
function

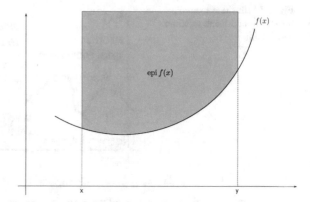

**Theorem 3.1** *If a mathematical program is convex then any local (i.e., relative)
minimum is a global minimum.*

The research subfield that deals with no convex programs are referred to as
global optimization, which aims at finding the globally best solution of models in
the potential presence of multiple local optima.

### 3.1.3   Duality

Duality is one of the most fundamental concepts in mathematical programming and
establishes a connection between two "symmetric" programs, namely, the primal
and dual problem. Duality is a powerful and widely employed tool in applied
mathematics for a number of reasons. First, the dual program is always convex even
if the primal is not. Second, the number of variables in the dual is equal to the
number of constraints in the primal that is often less than the number of variables in
the primal program. Third, the maximum value achieved by the dual problem is
often equal to the minimum of the primal.

The dual function is introduced as:

$$\xi(\lambda, \mu) = \underset{x}{\text{Infimun}} \{f(x) + \lambda^T \phi(x) + \mu^T \psi(x)\} \tag{3.2}$$

Then, the dual problem of the primal problem (3.1) is defined as follows:

$$\begin{aligned} \underset{\lambda, \mu}{\text{maximize}} \quad & \xi(\lambda, \mu) \\ \text{subject to} \quad & \mu \geq 0 \end{aligned} \tag{3.3}$$

Hence, using the Lagrange function, the dual problem can also be rewritten as:

$$\underset{\lambda, \mu; \mu \geq 0}{\text{maximize}} \left\{ \underset{x}{\text{Infimum}} \quad \Gamma(x, \lambda, \mu) \right\} \tag{3.4}$$

where the vectors $\lambda$ and $\mu$ are called Lagrange multipliers, and the Lagrange function is defined by

$$\Gamma(x, \lambda, \mu) = f(x) + \lambda^T h(x) + \mu^T g(x) \tag{3.5}$$

The Theorem 3.2 establishes an important relationship between the dual and primal problems.

**Theorem 3.2** Weak duality *For any feasible solution $x$ of the primal problem (3.1) and for any feasible solution $\lambda, \mu$, of the dual problem (3.3), the following holds*

$$f(x) \geq \xi(\lambda, \mu) \tag{3.6}$$

In addition, the theorem 3.3 is of relevant importance in mathematical programming. It shows that for convex programs, the primal problem solution can be obtained by solving the dual problem.

**Theorem 3.3** *If the primal problem is convex, then $f(x^*) = \xi(\lambda^*, \mu^*)$. Otherwise, one or both of the two sets of feasible solutions is empty.*
Note that $x^*$ represents the optimal solution of the primal problem, and $\lambda^*, \mu^*$ are the optimal solutions of the dual problem.

In nonconvex programs, there is a difference between the optimal objective function values of the dual and primal problems $(\xi(\lambda^*, \mu^*) - f(x^*))$, which is called *duality gap*. In convex programs, the duality gap is zero. According to Conejo at al. (2002), for nonconvex programs of engineering applications, the duality gap is usually relatively small.

## 3.2 Linear Programming

Linear Programming (LP) is a technique for the optimization of a linear objective function, subject to linear equality and/or linear inequality constraints. Given a polytope and a real-valued affine function defined on this polytope, a LP method will find a point on the polytope where this function has the optimal value if such point exists, by searching through the polytope vertices.
LP problems can be expressed in the standard form as follows:

$$\text{maximize}\{c^T x : Ax \leq b, x \geq 0\} \tag{3.7}$$

where $x$ represents the vector of decision variables (to be determined), $c$ and $b$ are vectors of (known) coefficients, and $A$ is a (known) matrix of coefficients. The expression to be optimized is called the objective function ($c^T x$ in this case). The equations $Ax \leq b$ are the constraints which specify a convex polytope over which the objective function is to be optimized.

The linear programming optimization and relevant solution algorithms such as Simplex and interior-point methods, are principally based on the following fundamental theorem:

**Theorem 3.4** *If an LP has an optimal solution; there is a vertex (i.e., extreme point) of the feasible polytope that is optimal.*

### 3.2.1   The Simplex Method

The Simplex algorithm, which was first developed by G. B. Dantzig in 1947, solves linear programs by moving along the boundaries from one vertex (extreme point) to the next. The algorithm starts with an initial vertex *basic feasible* solution and tests its optimality. The algorithm terminates, if some optimality condition is verified, otherwise, the algorithm identifies an adjacent vertex, with a better objective value. The optimality of this new solution is tested again, and the entire scheme is repeated, until an optimal vertex is finally found. Since every time a new vertex is identified the objective value is improved (except from a certain pathological case), and the set of vertices is finite, it follows that the algorithm will terminate in a finite number of iterations. Given the above description of the algorithm, it is inferred that the Simplex essentially starts from some initial extreme point, and follows a path along the edges of the feasible region toward an optimal extreme point, such that all the intermediate extreme points visited are not worsening the objective function (see Fig. 3.4a). It is worth mentioning that in 1953, Dantzig and Orchard-Hays proposed the Revised Simplex method, which actually is not a different method but is a different (more efficient) way to carry out each computational step of the Simplex method.

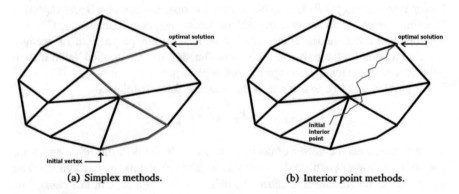

(a) Simplex methods.                    (b) Interior point methods.

**Fig. 3.4** Graphical interpretation of linear programming methods

### 3.2.2 Interior-Point Methods

During the period 1979–1996, there has been intensive interest in the development of interior-point methods. A theoretical breakthrough came in 1979 when L. G. Khachian discovered an ellipsoid algorithm whose running time in its worst case was significantly lower than that of the Simplex algorithm. Other theoretical results quickly followed, notably that of N. Karmarkar who discovered an interior-point algorithm whose running time performance in its worst case was significantly lower than that of Kachiyan's. This, in turn, was followed by more theoretical results by others improving on the worst-case performance. In a nutshell, an interior-point algorithm is one that improves a feasible interior solution point of the linear program by steps through the interior, rather than one that improves by steps around the boundary of the feasible region, as the Simplex algorithm does (see Fig. 3.4).

Assuming an initial feasible interior point is available and that all moves satisfy the whole set of constraints, the key ideas behind interior-point methods are as follows:

- Try to move through the interior in directions that show promise of moving quickly to the optimal solution.
- Recognize that if we move in a direction that sets the new point too "close" to the boundary, this will be an obstacle that will impede our moving quickly to an optimal solution. One way around this is to transform the feasible region so that the current feasible interior point is at the center of the transformed feasible region. Once a movement has been made, the new interior point is transformed back to the original space, and the whole process is repeated with the new point as the center.
- The simple stopping rule typically followed is to stop with an approximate optimal solution when the difference between iterates "deemed" sufficiently small in the original space.

The interested reader is referred to Dantzig and Thapa (1997, 2003) for a detailed description of the basic principles, the theory, and extensions of linear programming algorithms.

## 3.3 Mixed Integer Programming

Mathematical programs, which some of its (decision) variables are integer and/or binary, are called *mixed integer programs*. Integer variables appear when modeling indivisible entities, while a very common use of binary (0–1) variables is to represent binary choice. Consider an event that may or may not occur, and suppose that it is part of the problem to decide between these possibilities. In order to model such a dichotomy, a binary variable, which typically equals 1 if the event occurs otherwise is set to zero, can be used. The event itself can be almost anything.

Depending on the specific problem, the event may represent yes/no decisions, logical conditions, fixed costs,or piecewise linear functions.

(Linear) Mixed Integer Programming (MIP) problems can be expressed in the standard form as follows:

$$\text{Maximize } c^T x + hy$$
$$Ax + Gy \leq b$$
$$x \geq 0 \tag{3.8}$$
$$y \geq 0 \text{ and integer or binary}$$

where $x$ represents the vector of nonnegative variables, $y$ represents the vector of integer and/or binary variables, $c$ and $b$ are vectors of coefficients, and $A$ and $G$ are matrices of coefficients. In this case, the objective function is $c^T x + hy$.

Principally, there are three methodologies for solving this type of programs: the branch-and-bound, the cutting-plane, and the branch-and-cut methods. A brief description of those methods follows.

### 3.3.1  Branch-and-Bound Methods

The branch-and-bound method is the basic workhorse technique for solving integer and discrete programming problems. The idea of branch-and-bound was introduced by Land and Doig (1960), and actually is based on the observation that the enumeration of integer solutions has a tree structure. More specifically, the solution of a problem with a branch-and-bound algorithm is described as a search through a tree, wherein the root node corresponds to the relaxed original problem, and each other node corresponds to a subproblem of the original problem. In MIP problems, the branch-and-bound algorithm only branches on the integer variables, therefore the discussion can be restricted to a purely integer problem without loss of generality.

For instance, consider the complete enumeration of a MIP model having one integer variable $1 \leq x_1 \leq 3$, and two binary variables $x_2$ and $x_3$. Figure 3.5 illustrates the complete enumeration of all solutions for these variables, even those that might be infeasible due to other constraints on the model. The structure in Fig. 3.5 looks like a tree lying on its side with the root node on the left, and the leaf nodes on the right. The leaf nodes represent the actual enumerated complete solutions; so there are 12 of them.

For example, the node at the upper right represents the solution in which $x_1 = 1$, $x_2 = 0$, and $x_3 = 0$. The other nodes can be thought of as representing sets of possible solutions. For example, the root node represents all solutions that can be generated by growing the tree. Another intermediate bud node, e.g., the first node directly to the right of the root node, represents another subset of all of the possible solutions, in this case, all of the solutions in which $x_1 = 2$ and the other two variables can take any of their possible values. For any two directly connected

**Fig. 3.5** An illustrative
example of branch-and-bound
enumeration tree

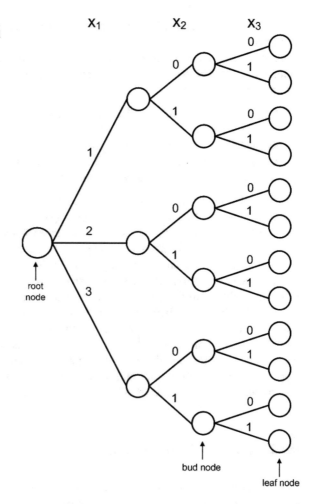

nodes in the tree, the parent node is the one closer to the root, and the child node is the one closer to the leaves.

The main idea in branch-and-bound method is to avoid growing the whole tree as much as possible, because the entire tree is just too big in any real problem. Instead branch-and-bound grows the tree in stages, and grows only the most promising nodes (i.e., partial or complete solutions) at any stage. It determines which node is the most promising by estimating a bound on the best value of the objective function that can be obtained by growing that node to later stages. The name of the method comes from the *branching* that happens when a bud node (i.e., partial solution, either feasible or infeasible) is selected for further growth and the next generation of children of that node is created. The *bounding* comes in when the bound on the best value attained by growing a node is estimated. Hopefully, in the end, branch-and-bound will have grown only a very small fraction of the full enumeration tree.

Additionally, the branch-and-bound algorithm attempts to reduce the amount of enumeration by *pruning* branches from the enumeration tree of all possible solution by applying two simple maxims:

- A branch can be eliminated (pruned), if it can be shown to contain no integer feasible solutions with a better value than the incumbent solution (i.e., the best complete feasible solution found so far).
- An upper bound for the integer solutions, on any branch, is always the relaxed. LP solution ignores the integer requirements.

The order in which the branch-and-bound algorithm proceeds after the first branch is governed by the *branching rules* adopted. Branching rules can range from *breadth-first*, which expand all possible branches from a tree node before going deeper in the tree, to *depth-first* that expand the deepest node first.

Finally, the branch-and-bound algorithm terminates when the incumbent solution's objective function value is better than or equal to the bounding function value associated with all of the bud nodes. This means that none of the bud nodes could possibly develop into a better solution than the complete feasible solution already has in hand, so there is no point in expanding the tree any further. Of course, according to the pruning policies, all bud nodes in this condition will already have been pruned, so this terminating rule amounts to saying that branch-and-bound stops when there are no more bud nodes left to consider for further growth. This also proves that the incumbent solution is optimum.

### 3.3.2   Cutting-Plane Methods

R. E. Gomory introduced cutting planes in the 1950s as a method for solving integer programming and MIP problems. However, most experts, including Gomory himself, considered them to be impractical due to numerical instability, as well as ineffective because many rounds of cuts were needed to make progress toward the solution. Things turned around in the mid-1990s when Conejo and co-workers showed them to be very effective in combination with branch-and-cut and ways to overcome numerical instabilities. Nowadays, all commercial MIP solvers use Gomory cuts in one way or another. Gomory cuts, however, are very efficiently generated from a simplex tableau, whereas many other types of cuts are either expensive or even NP-hard to separate. Among other general cuts for MIP, most notably lift-and-project dominates Gomory cuts. Other well-known cutting-plane methods include the Kelley's method and the Kelley–Cheney–Goldstein method.

The basic idea of cutting-plane methods is to alter the convex set of solutions to the related continuous LP problem (i.e., the LP problem that results by dropping the integer constraints) so that the optimal extreme point to the changed continuous problem is integer-valued. This is accomplished by systematically adding additional constraints (cutting planes) that cut off parts of the convex set that do not contain any feasible integer points and solving the resultant problems by the simplex algorithm. Note that an adding cut to a current fractional (i.e., not satisfying integrality) solution must assure that every feasible integer solution of the actual program is feasible for the cut, and the current fractional solution is not feasible for the cut.

### 3.3.3  Branch-and-Cut Methods

Branch-and-cut method is a hybrid of branch-and-bound and cutting-plane methods. The method solves the LP without the integer constraint using the regular simplex algorithm. When an optimal solution is obtained, and this solution has a non-integer value for a variable that is supposed to be an integer, a cutting-plane algorithm is used to find further linear constraints which are satisfied by all feasible integer points but violated by the current fractional solution. If such an inequality is found, it is added to the LP, such that resolving it will yield a different solution that is hopefully "less fractional". This process is repeated until either an integer solution is found (which is then known to be optimal) or until no more cutting planes are found.

At this point, the branch-and-bound part of the algorithm begins. The problem is split into two versions, one with the additional constraint that the variable is greater than or equal to the next integer greater than the intermediate result, and one where this variable is less than or equal to the next lesser integer. In this way, new variables are introduced in the basis according to the number of basic variables that are non-integers in the intermediate solution but which are integers according to the original constraints. The new LPs are then solved using the simplex method and the process repeats until a solution satisfying all the integer constraints is found. During the branch-and-bound process, further cutting planes can be separated, which may be either global cuts (i.e., valid for all feasible integer solutions) or local cuts (i.e., satisfied by all solutions fulfilling the side constraints from the currently considered branch-and-bound sub-tree).

### 3.3.4  Other Methods

It is worth mentioning a special set of integer programs called disjunctive programs. Roughly speaking, disjunctive programs comprise a logical system of conjunctive and disjunctive statements, where each statement is defined by a constraint. The basic theory of disjunctive programming can be found in the contribution works of Raman and Grossmann (1994) and Lee and Grossmann (2000). A distinctive methodology for solving mixed integer nonlinear programs is the outer approximation algorithm developed by Duran and Grossmann (1986). Finally, for more details about MIP algorithms, the reader is referred to Wolsey (1998), Nemhauser and Wolsey (1999) and Gass (2003).

## 3.4   Software

In this section, a brief description of the commercial software used to solve the optimization models is presented in this book. There exist a number of commercial tools for general modeling and optimization purposes such as GAMS, AIMMS, AMPL, and ILOG, which render very similar characteristics. In this book, GAMS has been used, since it is the most widely used modeling and optimization software in the PSE community. CPLEX solver has been selected for solving the MIP problems addressed throughout the book.

### 3.4.1   GAMS—General Algebraic Modeling System

GAMS was the first algebraic modeling language and is formally similar to commonly used fourth-generation programming languages. GAMS contains an integrated development environment (i.e., a language compiler) and is connected to a group of integrated high-performance third-party optimization solvers such as CPLEX, BARON, GUROBI, CONOPT, and XPRESS. GAMS is tailored for complex, large-scale modeling applications, and allows building large maintainable models that can be adapted quickly to new situations.

According to Rosenthal (2010) and Castillo et al. (2001), some of the more remarkable features of GAMS algebraic modeling language are:

- The model representation is analogous to the mathematical description of the problem. Therefore, learning GAMS programming language is almost natural for those working in the optimization field. Additionally, GAMS is formally similar to commonly used programming languages.
- Models are described in compact and concise algebraic statements that are easy for both humans and machines to read.
- The modeling task is completely apart from the solving procedure. Once the model of the system in question has been built, one can choose among the diverse solvers available to optimize the problem.
- Allows changes to be made in model specifications simply and safely.
- Allows unambiguous statements of algebraic relationships.
- Permits model descriptions that are independent of solution algorithms.
- All data transformations are specified concisely and algebraically. This means that all data can be entered in their most elemental form and that all

transformations made in constructing the model and in reporting are available for inspection.

- The ability to model small size problems and afterward to transform them into large-scale problems without significantly varying the code.
- Decomposition algorithms can be programmed in GAMS by using specific commands, thus not requiring additional software.
- GAMS imports/exports data from/to Microsoft EXCEL. Additionally, GAMS can be easily linked with MATLAB (The Mathworks 1998) using the Matgams library (Ferris 1999) if some special data manipulation is needed.

### 3.4.2 CPLEX Solver

IBM ILOG CPLEX, often informally referred to simply as CPLEX, is an optimization solver package. It is named for the Simplex method and the C programming language, although today it contains interior-point methods and interfaces in the C++, C#, and Java programming languages. GAMS/CPLEX is a GAMS solver that allows users to combine the high-level modeling capabilities of GAMS with the power of CPLEX optimizers. CPLEX optimizers are designed to solve large, difficult problems quickly and with minimal user intervention. Access is provided (subject to proper licensing) to CPLEX solution algorithms for linear, quadratically constrained, and mixed integer programming problems. While numerous solving options are available, GAMS/CPLEX automatically calculates and sets most options at the best values for specific problems. It is worth mentioning that for problems with integer variables, CPLEX uses a branch-and-cut algorithm that solves a series of LP subproblems. Because a single mixed integer problem generates many subproblems, even small MIP problems can be very compute intensive and require significant amounts of physical memory.

## 3.5  Final Remarks

In this chapter, the major optimization techniques and tools used throughout this book have been presented. The main concepts beneath each method have been briefly described in order to provide the reader with a general understanding of the theory involved in the solution approaches.

Indeed, the mathematical model of a system is the collection of mathematical relationships which—for the purpose of developing a design or plan—characterizes the set of feasible solutions of the system. Precisely, being the scope of this book the development and presentation of techniques for the efficient solution of large-scale production scheduling and planning problems in the process industries, it becomes of utmost importance the discovery of tailored strategies for specific industrial sectors that conform specific MIP modeling techniques (Chaps. 4, 5, 6, and 7) and MIP-based specific solution approaches (decomposition techniques in Chap. 8). At this point, it is worth noticing that *the process of building a mathematical model is often considered to be as important as solving it* because this process provides insight about how the system works and helps organize essential information about it. Models of the real world are not always easy to formulate because of the richness, variety, and ambiguity that exists in the real world or because of our ambiguous understanding of it. As a result, building up concise, useful and efficient mathematical models/approaches is a very difficult and challenging task.

# References

Castillo E, Conejo AJ et al (2001) Building and solving mathematical programming models in engineering and science. Wiley, New York, USA

Conejo AJ, Nogales FJ, Prieto FJ (2002) A decomposition procedure based on approximate Newton directions. Math Program 93(3):495–515

Dantzig GB, Thapa MN (1997) Linear programming 1: introduction. Series in operations research. Springer-Verlag New York, Inc., USA

Dantzig GB, Thapa MN. (2003) Linear programming 2: theory and extensions. Series in operations research. Springer-Verlag New York, Inc., USA

Duran M, Grossmann IE (1986) An outer-approximation algorithm for a class of mixed integer nonlinear programs. Math Progr 36:307–339

Ferris, M. MATLAB and GAMS (1999): Interfacing optimization and visualization software. Technical report, University of Wisconsin

Gass SI (2003) Linear programming: methods and applications. Dover Publications, Inc., Mineola, New York, USA

Land AH, Doig AG (1960) An automatic method for solving discrete programming problems. Econometrica 28:497–520

Lee S, Grossmann IE (2000) New algorithms for nonlinear disjunctive programming. Comput Chem Eng 24:2125–2141

Nemhauser G, Wolsey LA (1999) Integer and combinatorial optimization. series in discrete mathematics and optimization. Wiley-Interscience, Wiley-Interscience, New York, USA

Raman R, Grossmann IE (1994) Modeling and computational techniques for logic based integer programming. Comput Chem Eng 18:563–578

Rosenthal RE (2010) GAMS—a user's guide. GAMS Development Corporation, Washington, DC, USA

The Mathworks, Inc. MATLAB (1998) The language of technical computing: MATLAB notebook user's guide. The Mathworks Inc., Natick, MA, USA

Wolsey LA (1998) Integer programming. Series in discrete mathematics and optimization. Wiley-Interscience, NewYork, USA

# Part II
# Continuous Processes

# Chapter 4
# Production Planning and Scheduling of Parallel Continuous Processes

## 4.1 Introduction

In this chapter, we focus on production processes with continuous parallel units, often also termed as single-stage continuous processes. Given the prominence of this class of problems, a number of stand-alone scheduling as well as integrated production planning and scheduling approaches have been proposed in the literature for similar problems, though no methods have been reported for the specific problem discussed here. Specifically, we developed a mathematical approach to the simultaneous production planning and scheduling of continuous parallel units producing a large number of final products that can be classified into product families. The problem under consideration appears in many stages of operation in process industries, including packing in batch and continuous production facilities. Thus, it is quite important since it arises in a number of different production environments (e.g., food and beverage industry, consumer products, etc.).

In contrast with previous research works, a more general case has been considered based on (i) product families, (ii) short planning periods that may lead to idle units for entire periods, (iii) changeovers spanning multiple periods, and (iv) maintenance activities. The motivation to consider product families comes from the fact that in many production environments, there exist products that share many characteristics. In fact, the current work was first developed to address problems in a highly complex real-life bottling facility producing hundreds of final products. The grouping into families is based on various criteria, including product similarities, processing similarities, or changeover considerations. The goal of the aforementioned grouping is to lead to computationally tractable optimization models without compromising the quality of solution. Furthermore, the use of product families

© Springer Nature Switzerland AG 2019
G. M. Kopanos and L. Puigjaner, *Solving Large-Scale Production Scheduling and Planning in the Process Industries*,
https://doi.org/10.1007/978-3-030-01183-3_4

reflects managerial practice prevalent in many production systems (Inman and Jones 1993; Grunow et al. 2002; Günther et al. 2006; Kopanos et al. 2010).

## 4.2   Problem Statement

The production planning and scheduling of continuous parallel units are considered here. The problem is defined in terms of the following items:

(i)    A known planning horizon that is divided into a set of periods, $n \in N$.
(ii)   A set of parallel processing units, $j \in J$, with available production time in period $n$ equal to $\omega_{jn}$.
(iii)  A set of product families or simply *families*, $f \in F$, wherein all products are grouped into; $f \in F_j$ is the subset of families that can be assigned to unit $j$, and $j \in J_f$ is the subset of units that can process family $f$.
(iv)   A set of products $i \in I$ with demand $\varsigma_{in}$ at the end of time period $n$, backlog $\psi_{in}$ and inventory $\xi_{in}$ costs, minimum and maximum production rates $\left( \rho_{ijn}^{min}, \rho_{ijn}^{max} \right)$, minimum processing times $\tau_{ijn}^{min}$, and processing cost $\lambda_{ij}$; the subset of products in family $f$ is denoted by $i \in I_f$; $i \in I_j$ is the subset of products that can be assigned to unit $j$, and $j \in J_i$ is the subset of units that can produce product $i$.
(v)    A sequence-dependent switchover operation, or simply *changeover*, is required on each processing unit whenever the production is changed between two different product families; the required changeover time is $\gamma_{ff'j}$, while the changeover cost is $\varphi_{ff'j}$; and
(vi)   A sequence-independent switchover operation, henceforth referred to as *setup*, is required whenever a product $i$ is assigned to a processing unit $j$; the setup time is $\delta_{ij}$ and the setup cost is $\theta_{ij}$.

We assume a non-preemptive operation mode, no utility restrictions, and that changeover times are not greater than a planning period.

The objective is to determine:

(i)    the assignment $\left( Y_{fjn}^F \right)$ of product families on each unit in every period;
(ii)   the sequencing $\left( X_{ff'jn} \right)$ between families on each unit in every period;
(iii)  the assignment of products to processing units in every period $\left( Y_{ijn} \right)$;
(iv)   the production amount for every product in every period $\left( P_{in} \right)$; and,
(v)    the inventory $\left( S_{in} \right)$ and backlog $\left( B_{in} \right)$ profiles for all products.

So as to satisfy customer demand at the minimum total cost, including operating, changeover and setup costs, as well as inventory and backlog costs.

## 4.3  Proposed Approach

The novelty of the proposed formulation lies in the integration of three different modeling approaches. In particular, we use (a) a discrete-time approach for inventory and backlog costs calculation for production planning, (b) a continuous-time approach with sequencing using immediate precedence variables for the scheduling of families, and (c) lot-sizing type of capacity constraints for the scheduling of products. Figure 4.1 illustrates the proposed modeling approach.

**Fig. 4.1** Proposed modeling approach (Kopanos et al. 2011)

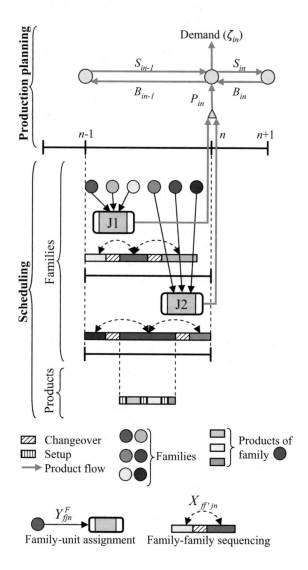

For the production planning subproblem, we employ a time grid with fixed, though not necessarily equal, production periods. The planning horizon is therefore divided into $n \in N = \{1, 2, \ldots\}$ periods; period $n$ starts at time point $n - 1$ and finishes at time point $n$. Note that the use of a discrete-time approach at the planning level (big-bucket) enables the correct calculation of holding and backlog costs. Material balances for each product are expressed at the end of each planning period in terms of total production level, $P_{in}$, inventory level, $S_{in}$, and backlog level, $B_{in}$. We assume that the amounts produced during a planning period become available at the end of this period. The *communication* between the production planning and scheduling subproblems is accomplished via the amount $Q_{ijn}$ of product $i$ produced in unit $j$ during period $n$. Variable $Q_{ijn}$ is used by the production planning problem for the calculation of variables $P_{in}$ for the material balances, while at the same time variables $Q_{ijn}$ is subject to detailed sequencing and capacity constraints of the scheduling subproblem.

The scheduling subproblem has two levels. At the first level, we schedule product families on units using an immediate precedence approach for sequencing, while at the second level, we employ a lot-sizing-based approach to express capacity constraints for products. In particular, we define assignment binary variable $Y_{fjn}^F$ to denote the assignment of family $f$ in unit $j$ during period $n$, and sequencing binary variable $X_{ff'jn}$ $(\bar{X}_{f'fjn})$ to denote an immediate precedence $f \rightarrow f'$ in unit $j$ within period $n$ (across periods $n - 1$ and $n$). This allows us to account for changeover times between families and correctly calculate changeover costs. The activation of the sequencing variables is achieved using a modification of the network formulation of Karmarkar and Schrage (1985) and Sahinidis and Grossmann (1991). For the timing of processing of family $f$ in unit $j$ during period $n$, we introduce variable $C_{fjn}$. Individual product setups and capacity constraints are modeled using lot-sizing-type setup binary variables: $Y_{ijn} = 1$ if product $i$ is produced in unit $j$ during period $n$. Variables $Y_{ijn}$ are used to constrain the processing time $T_{ijn}$, which is in turn used to constrain the production amount $Q_{ijn}$.

In addition to this novel integration of modeling approaches, our formulation can accurately account for changeovers in the presence of fixed planning periods. First, it allows us to track the last family produced in each period, and therefore account correctly for the first changeover in the following period. If the last family in period $n - 1$ and the first in period $n$ are the same, then no changeover time/cost is added (changeover carryover). Second, it allows changeover operations to cross planning period boundaries (changeover crossover), thereby allowing higher utilization of resources and obtaining better solutions (see Sung and Maravelias (2008) for a discussion of these aspects in the context of lot-sizing problems). Finally, an extension of our approach allows the seamless modeling of idle planning periods, another aspect that has often been neglected, though it can appear in optimal solutions when the length of the planning period is short and/or a unit is not heavily loaded.

## 4.4  Mathematical Formulation

In this section, we present a MIP formulation for the production planning and scheduling of a parallel unit (single-stage) facility described in Sect. 4.2. Constraints are grouped according to the type of decision (e.g., assignment, timing, sequencing, etc.). To facilitate the presentation of the model, we use uppercase Latin letters for optimization variables and sets and lowercase Greek letters for parameters.

**Material Balance Constraints** The total amount $P_{in}$ of product $i$ produced in period $n$ is the summation of the produced quantities $Q_{ijn}$:

$$P_{in} = \sum_{j \in J_i} Q_{ijn} \quad \forall i, n \tag{4.1}$$

Mass balances for every product $i$ are expressed at the end of each production period $n$, where initial backlogs ($B_{in} = 0$) and initial inventories ($S_{in} = 0$) are given by

$$S_{in} - B_{in} = S_{in-1} - B_{in-1} + P_{in} - \varsigma_{in} \quad \forall i, n \tag{4.2}$$

Inventory capacity constraints can be enforced via constraints similar to:

$$S_{in} \leq \text{product storage capacity} \quad \forall i, n \quad \text{or}$$
$$\sum_i S_{in} \leq \text{plant storage capacity} \quad \forall n \tag{4.3}$$

**Family Allocation Constraints** Obviously, a family $f$ is assigned to a processing unit $j \in J_f$ during a production period $n$ if at least one product that belongs to this family, $i \in I_f$ is produced on this processing unit at the same period:

$$Y^F_{fjn} \geq Y_{ijn} \quad \forall f, i \in I_f, j \in J_f, n \tag{4.4}$$

where $Y^F_{fjn}$ denotes a family-to-unit assignment, and $Y_{ijn}$ denotes a product-to unit assignment. Moreover, constraint set (4.5) enforces the binary $Y^F_{fjn}$ to be zero when no product $i \in I_f$ is produced on unit $j$ in period $n$:

$$Y^F_{fjn} \leq \sum_{i \in I_f} Y_{ijn} \quad \forall f, j \in J_f, n \tag{4.5}$$

Note that if we do not include constraint set (4.5), we may be obtaining solutions where $Y^F_{fjn} = 1$ for a family that $\sum_{i \in I_f} Y_{ijn} = 0$; that also means $T^F_{fjn} = 0$. Note that this undesired case could be observed if the changeover cost/time for $f \to f'$ is higher than the sum of costs/times for changeovers $f \to f''$ and $f'' \to f'$.

**Family Sequencing and Timing Constraints** We introduce binary variable $X_{ff'jn}$ to define the local precedence between two families $f$ and $f'$: $X_{ff'jn} = 1$ if family $f'$ is processed immediately after family $f$ in unit $j$. Constraints (4.6) and (4.7) ensure that, if family $f$ is allocated on the processing unit $j$ at period $n$ (i.e., $Y_{fjn}^F = 1$), at most one family $f'$ is processed before and after it, respectively. If family $f$ is assigned first on a processing unit $j$ (i.e., $WF_{fjn} = 1$), then it has no predecessor. Similarly, if family $f$ is assigned last on a processing unit $j$ (i.e., $WL_{fjn} = 1$), then has no successor.

$$\sum_{f' \neq f, f' \in F_j} X_{f'fjn} + WF_{fjn} = Y_{fjn}^F \quad \forall f, j \in J_{f,n} \tag{4.6}$$

$$\sum_{f' \neq f, f' \in F_j} X_{ff'jn} + WL_{fjn} = Y_{fjn}^F \quad \forall f, j \in J_{f,n} \tag{4.7}$$

The correct number of immediate precedence variables is activated through constraint set (4.8), which enforces the total number of sequenced pairs within a period to be equal to the total number of active assignments during this period minus one:

$$\sum_{f \in F_j} \sum_{f' \neq f, f' \in F_j} X_{ff'} + 1 = \sum_{f \in F_j} Y_{fjn}^F \quad \forall j, n \tag{4.8}$$

To avoid sequence subcycles, we also include constraint set (4.9), which ensures a feasible timing of the families assigned to the same processing unit:

$$C_{f'jn} \geq C_{fjn} + T_{f'jn}^F + \gamma_{ff'j} X_{ff'jn} - \omega_{jn}(1 - X_{ff'jn})$$
$$\forall f, f' \neq f, j \in (J_f \cap J_{f'}), n \tag{4.9}$$

**Family Changeovers Across Adjacent Periods** We introduce binary variable $\bar{X}_{ff'jn}$ to denote a changeover from family $f$ to $f'$ in unit $j$ taking place at the beginning of period $n$. This binary variable is active only for the family $f$ processed last in period $n - 1$ (i.e., $WL_{fjn-1=1}$) and the family $f$ 0 that is processed first in period $n$ (i.e., $WF_{f'jn} = 1$), according to constraints (4.10) and (4.11) (see Fig. 4.2).

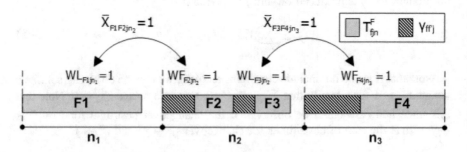

Fig. 4.2 Family changeovers between adjacent periods

$$WF_{fjn} = \sum_{f' \in F_j} \bar{X}_{f'fjn} \quad \forall f, j \in J_{f}, n > 1 \tag{4.10}$$

$$WL_{fjn-1} = \sum_{f' \in F_j} \bar{X}_{ff'jn} \quad \forall f, j \in J_{f}, n > 1 \tag{4.11}$$

**Changeover Crossover Constraints** In most existing approaches, changeover times have to begin and finish within the same period. In other words, crossovers of changeover times are not allowed. This restriction may result in suboptimal solutions. For example, in Fig. 4.3, a better solution can be obtained if the changeover from F1 to F2 starts in period $n_1$ and finishes in $n_2$.

To model changeover crossovers, nonnegative variables $\bar{U}_{jn}$ and $U_{jn}$ are introduced. If a changeover operation with a duration equal to $\gamma_{ff'j}$ starts in period $n - 1$ and is continued in period $n$, then: $U_{jn-1}$ represents the fraction of time of the changeover operation that takes place in period $n - 1$; and $\bar{U}_{jn}$ represents the fraction of time of the operation that is performed in period $n$ (see Fig. 4.4).

$$\bar{U}_{jn} + U_{jn-1} = \sum_{f \in F_j} \sum_{f' \in F_j, f' \neq f} \gamma_{ff'j} \bar{X}_{ff'jn} \quad \forall j, n > 1 \tag{4.12}$$

**Unit Production Time** The summation of the production times of families $f \in F_j$ that are processed on unit $j$ plus the total changeover times within period $n$ (including variables $\bar{U}_{jn}$ and $U_{jn}$) is constrained by the available production time $\omega_{jn}$:

$$\bar{U}_{jn} + U_{jn} + \sum_{f \in F_j} T^F_{fjn} + \sum_{f \in F_j} \sum_{f' \in F_j, f' \neq f} \gamma_{ff'j} X_{ff'jn} \leq \omega_{jn} \quad \forall j, n \tag{4.13}$$

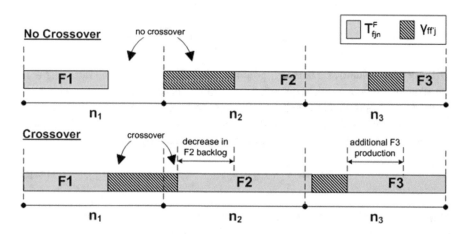

**Fig. 4.3** Crossover of changeover times

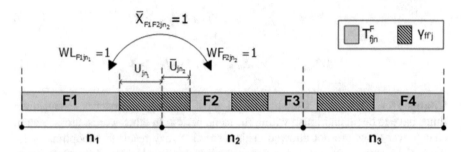

**Fig. 4.4** Modeling of crossover of changeover times

where $\bar{U}_{jn=1}$ corresponds to the time point that unit $j$ is available to begin processing any task in the first period. Notice that $\bar{X}_{ff'jn=1} = 0$ by definition.

**Product Lot-Sizing Constraints** Upper and lower bounds on the production, $Q_{ijn}$, of product $i$ on unit $j$ during period $n$ are enforced by constraint (4.14), where variable $T_{ijn}$ denotes the processing time of product $i$ on unit $j$ in period $n$,

$$\rho_{ij}^{\min} T_{ijn} \le Q_{ijn} \le \rho_{ij}^{\max} T_{ijn} \quad \forall i,\, j \in J_{i,n}. \tag{4.14}$$

Upper and lower bounds on the processing time $T_{ijn}$ are enforced by

$$\tau_{ijn}^{\min} Y_{ijn} \le T_{ijn} \le \tau_{ijn}^{\max} Y_{ijn} \quad \forall i,\, j \in J_{i,n} \tag{4.15}$$

Note that a tight processing time upper bound $\tau_{ijn}^{\max}$ can be estimated as follows:

$$\tau_{ijn}^{\max} = \omega_{jn} \qquad\qquad \text{if } \sum_{n'-1}^{N} \zeta_{in'}/\rho_{ij}^{\min} \ge \omega_{jn},$$

$$\tau_{ijn}^{\max} = \sum_{n'-1}^{N} \zeta_{in'}/\rho_{ij}^{\min} \quad \text{if } \sum_{n'-1}^{N} \zeta_{in'}/\rho_{ij}^{\min} < \omega_{jn}.$$

Since, the switchovers between products that belong to the same family $(i \in I_f)$ are sequence-independent, sequencing and timing decisions regarding products can be made post optimization without affecting the quality of solution. For example, the sequencing can be determined depending on the characteristics of the production process. The family processing time $T_{fjn}^{F}$ is defined by (see Fig. 4.5).

$$T_{fjn}^{F} = \sum_{i \in I_f} (T_{ijn} + \delta_{ij} Y_{ijn}) \quad \forall f,\, j \in J_{f,n} \tag{4.16}$$

**Objective Function** The optimization goal is the minimization of total inventory, backlog, changeover (inside and across periods), setup, and operating costs:

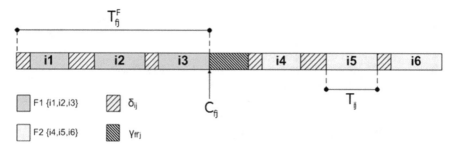

**Fig. 4.5** Family and product processing times

$$\min \sum_i \sum_n (\xi_{in} S_{in} + \psi_{in} B_{in})$$

$$+ \sum_f \sum_{f' \neq f} \sum_{j \in (J_f \cap J_{f'})} \sum_n \phi_{ff'jn} + (X_{ff'jn} + \bar{X}_{ff'jn}) \tag{4.17}$$

$$+ \sum_i \sum_{j \in J_i} \sum_n (\theta_{ij} Y_{ijn} + \lambda_{ij} Q_{ijn})$$

**Integrality and Nonegativity Constraints** . The domains of decision variables are defined as follows:

$$Y_{ijn} \in \{0, 1\} \quad \text{and} \quad Q_{ijn}, \quad T_{ijn} \geq 0 \quad \forall i, j \in J_{i,n}$$

$$WF_{fjn}, WL_{fjn}, Y^F_{fjn} \in \{0, 1\} \quad \text{and} \quad T^F_{fjn}, C_{fjn} \geq 0 \quad \forall f, j \in J_{f,n}$$

$$X_{ff'jn}, \bar{X}_{ff'jn} \in \{0, 1\} \quad \forall f, f', j \in (J_f \cap J'_f), n \tag{4.18}$$

$$B_{in}, S_{in}, P_{in} \geq 0 \quad \forall i, n$$

$$U_{jn}, \bar{U}_{jn} \geq 0 \quad \forall i, n$$

The proposed MIP model, CR, consists of constraints (4.1)–(4.18).

**Extension I: Idle Units** Note that model CR, similarly to most existing approaches, is based on the assumption that units do not remain completely idle in any period. In other words, processing units produce at least one product in each period, except maintenance periods. Generally speaking, this assumption is valid for medium to long planning periods (e.g., a production week). However, if short periods are used (e.g., a production day) to accurately model frequent intermediate due dates, idle periods may be present in an optimal solution.

To model unit idle periods, we define a dummy product $i \in I^{idle}_f$ for each family, with zero setup time and cost (i.e., $\delta_{ij} = 0$, $\theta_{ij} = 0$ $\forall i \in I^{idle}_f$). The processing times of dummy products are then constrained by (4.15) with $\tau^{min}_{ijn} = 0$ and $\tau^{max}_{ijn} = \omega_{jn}$. Note that if a processing unit produces only a dummy product in a production period then this actually means that the unit remains idle during that period. Having defined a variable for idle time, we can now express constraint (4.13) as equality,

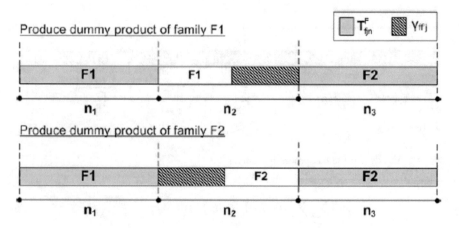

**Fig. 4.6** Modeling of changeover crossover through idle periods

where $T_{ijn} \forall i \in I_f^{\text{idle}}$ now act as slack variables. The new MIP model for idle units is named CR-D.

Figure 4.6 shows an illustrative example of a single-unit production plan over three production periods, where the unit produces family F1 in period $n_1$, remains idle during period $n_2$, and produces family F2 in period $n_3$ (white boxes denote the imaginary production of dummy product $i \in I_f^{\text{idle}}$). Note that the two solutions are equivalent.

**Extension II: Maintenance Activities** Maintenance activities in a given period can be readily addressed by fixing the corresponding changeover crossover variables to zero and modifying the available production time $\omega_{jn}$ accordingly. We assume that the maintenance activity is carried out at the end of the production period $n$. Our model can accommodate cases where the duration of a maintenance task is equal to the available production time (by setting $\omega_{jn} = 0$) or cases where the duration of the maintenance activity is smaller than the length of the period $n$. Note that if a maintenance task is performed between the production of two different families, there is no need for a changeover operation. Figure 4.7 illustrates our approach. Finally, note that we can also cope with maintenance tasks whose duration is greater than a planning period.

## 4.5  Applications

In this section, we discuss the application of the proposed model to an illustrative example and a large-scale industrial case study. All problem formulations were solved on a Sun Ultra 4.0 Workstation with 8 GB RAM using CPLEX 11 via a GAMS 22.9 interface. A maximum time limit of 300 CPUs was used. It represents

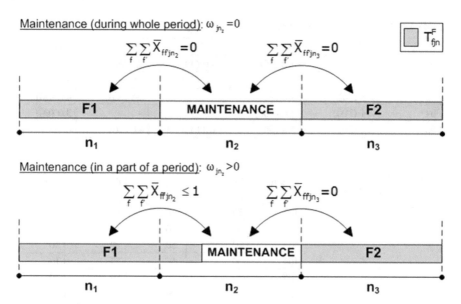

**Fig. 4.7** Modeling of maintenance activities

the amount of time practitioners are willing to wait for a solution. This is because different scenarios have to be tested before a solution is dispatched, and rescheduling solutions should be generated routinely and in a timely fashion. Furthermore, we selected this short time limit to make a fair comparison of our method to commercial tools that typically yield solutions within a few minutes.

### 4.5.1  Illustrative Example

We consider a simple example of 15 products (I01–I15), grouped in 5 families (F01–F05), and 3 processing units (J01–J03). All products can be produced on any unit. Products are grouped into families as follows: $I_{F01} = \{I01,\ I02,\ I03\}$, $I_{F02} = \{I04,\ I05,\ I06\}$, $I_{F03} = \{I07,\ I08,\ I09\}$, $I_{F04} = \{I10,\ I11,\ I12\}$, and $I_{F05} = \{I13,\ I14,\ I15\}$. The total production horizon is 4 days and is divided into four 24-h periods. Maintenance is scheduled on units J01, J02, and J03, in periods $n_2$, $n_3$, and $n_4$, respectively. Processing data includes setup time $\delta_{ij} = 0.5$; setup cost $\theta_{ij} = 50$; operating cost $\lambda_{ij} = 0.1$; production rate $\rho_{ij}^{\max} = 10$; minimum processing time $\tau_{ij}^{\min} = 0.2$; inventory cost $\xi_{in} = 1$; and backlog cost $\psi_{in} = 3$ for all products. Changeover times $\gamma_{ff'j}$ and costs $\phi_{ff'j}$ between product families are given in Table 4.1. Product demands can be found in Table 4.2. The processing sequence of

**Table 4.1** Illustrative example: changeover times (costs)

| Family | F01 | F02 | F03 | F04 | F05 |
|--------|-----|-----|-----|-----|-----|
| F01 | – | 3.0 (50) | 3.0 (40) | 5.0 (60) | 1.5 (50) |
| F02 | 5.3 (40) | – | 3.0 (50) | 3.0 (80) | 2.0 (90) |
| F03 | 2.8 (70) | 4.0 (30) | – | 2.5 (80) | 4.0 (30) |
| F04 | 2.4 (100) | 4.0 (100) | 3.0 (90) | – | 3.0 (60) |
| F05 | 3.2 (30) | 4.0 (50) | 2.0 (50) | 4.0 (70) | – |

**Table 4.2** Illustrative example: product demands per period (kg)

| Product | Day 1 $(n_1)$ | Day 2 $(n_2)$ | Day 3 $(n_3)$ | Day 4 $(n_4)$ |
|---------|------|------|------|------|
| I01 | 50 | 0 | 70 | 20 |
| I02 | 0 | 80 | 10 | 50 |
| I03 | 30 | 50 | 20 | 30 |
| I04 | 0 | 10 | 75 | 10 |
| I05 | 70 | 90 | 10 | 20 |
| I06 | 65 | 0 | 75 | 0 |
| I07 | 40 | 50 | 0 | 30 |
| I08 | 0 | 45 | 0 | 0 |
| I09 | 55 | 0 | 45 | 15 |
| I10 | 10 | 100 | 30 | 50 |
| I11 | 40 | 15 | 20 | 30 |
| I12 | 0 | 95 | 40 | 30 |
| I13 | 80 | 0 | 40 | 30 |
| I14 | 0 | 50 | 0 | 0 |
| I15 | 0 | 0 | 0 | 60 |

products that belong to the same family during every period $n$ is predetermined; specifically, it is in ascending index order.

The optimization goal is the minimization of the total cost, as defined in Eq. (4.17). The proposed MIP model obtained the optimal solution ($2630) in 30 CPUs (see Table 4.3). The optimal production schedule for products and product families is shown in Fig. 4.8. Notice the changeover crossover between family F02 and F01 on unit J01 across the boundary between the third and fourth day. Figure 4.9 presents the production profiles for product families and products, while Fig. 4.10 shows the inventory and backlog profiles for every product. High inventories are observed in the first day; especially for products I05 (90 kg), I07 (50 kg), and I13 (40 kg). High backlogs are observed in the second day, for products I08 (45 kg) and I04 (10 kg), and in the third day, for products I10 (30 kg), I11 (20 kg), and I12 (15 kg).

**Table 4.3** Computational results for all problem instances

| Problem instance | Constraints | Continuous variables | Binary variables | Nodes | Objective function | CPUs |
|---|---|---|---|---|---|---|
| Ilustrative example II | 1147 | 708 | 510 | 10,501 | 2630 | 30 |
| Case study: Instance I | 5192 | 3730 | 1890 | 97,017 | 235,577 | 1193 |
| Case Instance: II | 8882 | 6269 | 3264 | 59,573 | 155,629 | 3300 |

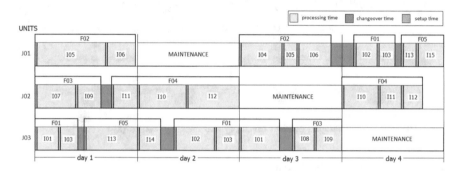

**Fig. 4.8** Illustrative example: Gantt chart of optimal solution

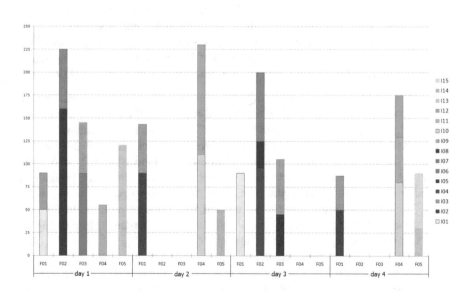

**Fig. 4.9** Illustrative example: production profiles of families and products (kg)

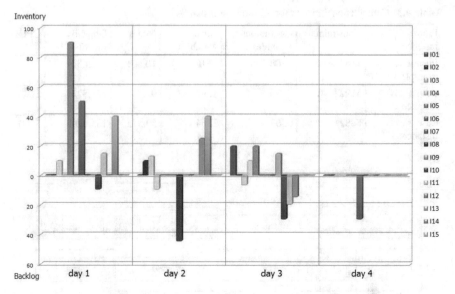

**Fig. 4.10** Illustrative example: product inventory and backlog profiles (kg)

## 4.5.2  *Industrial Case Study*

In this subsection, we consider a complex real-world problem in the continuous bottling stage of the Cervecería Cuauhtémoc Moctezuma beer production facility, situated in Mexico. The facility under study consists of 8t processing units (J01–J08), working in parallel and producing a total of 162 products that are grouped into 22 families (F01–F22). Process data are not provided due to confidentiality issues.

**Instance I** In this instance, we study the planning-scheduling problem over a 6-week production horizon divided into six 168-h periods. The problem was solved to optimality using model CR in less than 200 CPUs. Model and solution statistics can be found in Table 4.3. Figure 4.11 shows the Gantt chart of the optimal solution ($235,577) for product families. Note that each family block is subdivided into a number of smaller product blocks separated by setup times. The production profiles of all families in the eight units of the facility are presented in Fig. 4.12, while Fig. 4.13 shows the total inventory and backlog cost profiles. In the first week, we observe a high inventory cost, while backlog cost remains low during all periods. Inventories represent 34% of the total cost.

Furthermore, the solution obtained using the approach presented in this study was compared against the solution that was found, dispatched, and executed in practice using a combination of in-house and commercial tools. The main cost components of the two solutions are compared in Fig. 4.14. Clearly, our solution is substantially better. In particular, inventory cost is 78% lower than the inventory cost of the executed solution, and backlog cost is less than 5% of the executed

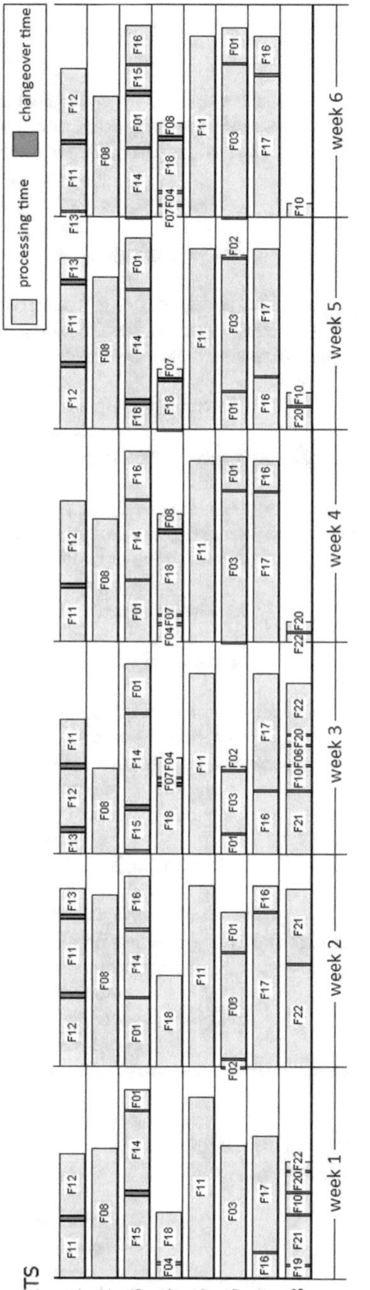

**Fig. 4.11** Industrial case study—Instance I: family Gantt chart of optimal solution

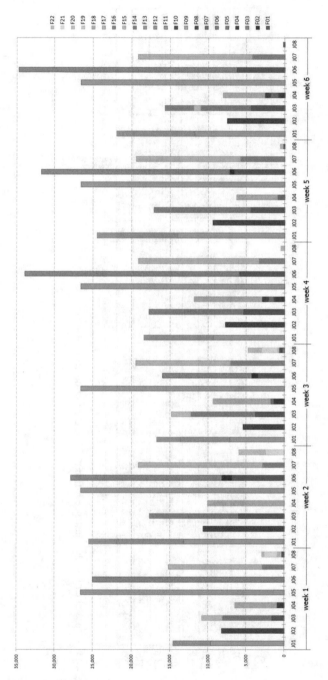

**Fig. 4.12** Industrial case study—Instance I: production profile for families (kg)

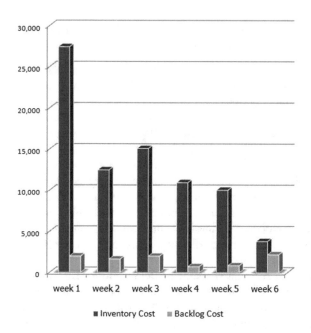

**Fig. 4.13** Industrial case study—Instance I: total inventory and backlog cost profiles ($)

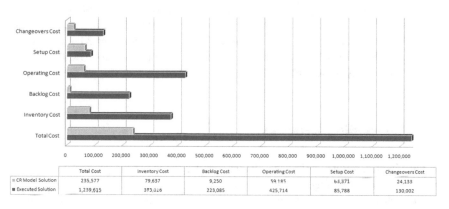

|  | Total Cost | Inventory Cost | Backlog Cost | Operating Cost | Setup Cost | Changeovers Cost |
|---|---|---|---|---|---|---|
| ▣ CR Model Solution | 235,577 | 79,637 | 9,250 | 59,185 | 63,371 | 24,133 |
| ▪ Executed Solution | 1,239,615 | 373,026 | 223,085 | 425,714 | 85,788 | 130,002 |

**Fig. 4.14** Industrial case study—Instance I: comparison of solutions obtained by the proposed MIP model and the tools currently used in practice ($)

solution. Also, changeover and setup costs have been reduced by more than 80 and 25%, respectively (see the table in the bottom of Fig. 4.14). The results of this case study indicate that the proposed framework can, in fact, be used to address real-world problems. It is computationally efficient and yields solutions of very good quality. Given the significant improvement over the practiced methods, our formulation is currently incorporated into and tested using the tools currently employed to generate detailed production plans.

**Instance II** In this instance, we discuss a problem on the same processing facility but over a planning horizon of 4 weeks partitioned into seven 24-h (week 1) and three 168-h planning periods (weeks 2–4). In other words, we consider daily orders during the first week, and weekly orders during the next 3 weeks. This partitioning represents industrial practice, where orders in the near future are known with certainty and are treated separately, while orders in later periods are aggregated. The consideration of such small planning periods leads to larger formulations that are harder to solve to optimality, but are necessary to avoid unmet demand. Furthermore, good solutions to these problems often require changeover crossovers, as well as idling of units for more than one (short) planning period. To address this instance, we developed a formulation that combines model CR-D for the first 7 days, allowing units to remain idle over a day; and model CR for the three 7-day production periods.

The model and solution statistics of the resulting MIP model can be found in Table 4.3. The best solution obtained within 300 CPUs has a total cost of $155,629 and has an optimality gap equal to 1%. The solution obtained using the proposed framework is again significantly better than the solution found by commercial tools. The production profile of all families is shown in Fig. 4.15, while a cost analysis for every week can be found in Fig. 4.16, where for week 1, we present the aggregated costs for day periods 1–7. It is worth noting that the consideration of 24-h planning periods during the first week results into higher inventory and backlog costs because daily demands are harder to meet. Inventories represent 31% of the total cost (vs. 34% in Instance I), but backlog costs have increased to 10% from 4% in Instance I. Finally, note that we do observe idle 24-h periods during the first week.

## 4.6  Concluding Remarks

In this chapter, a novel MIP formulation was presented for the production planning and scheduling of single-stage continuous processes with product families, a type of production facility that appears in a number of different production environments and industrial sectors. Our approach combines a discrete-time partitioning of the planning horizon to account for the major planning decisions (production targets, shipments, and inventory levels) with a continuous-time treatment of detailed scheduling decisions within each planning period. Furthermore, at the scheduling level, it combines a *precedence-based* approach to correctly enforce sequencing constraints among product families with a *lot-sizing-like* approach to account for production time constraints for individual products. Our approach addresses appropriately aspects such as changeover carryover and crossover, thereby leading to solutions with higher utilization of resources. Also, it is not based upon the assumption that processing units cannot remain idle during a production period,

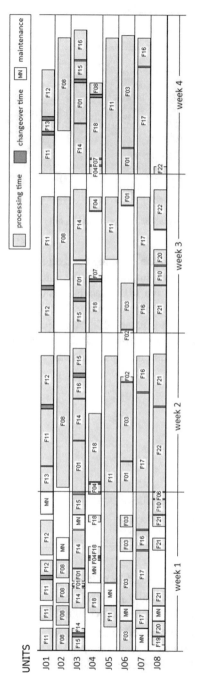

**Fig. 4.15** Industrial case study—Instance II: family Gantt chart of best solution

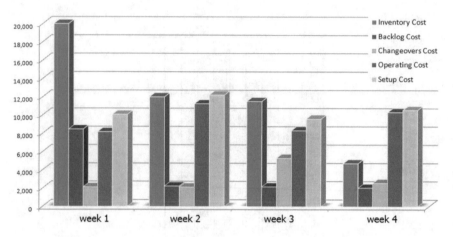

**Fig. 4.16** Industrial case study—Instance II: cost analysis ($)

thereby allowing us to partition the planning horizon into smaller periods, which in turn results in better solutions. Importantly, the integration of these approaches leads to computationally effective MIP models. Very good solutions to problems with hundreds of products can be obtained within 5 CPU min, while optimal solutions can also be found in a reasonable time. Furthermore, the proposed formulation yields solutions that are substantially better than the ones obtained using commercial tools, suggesting that MIP methods can be used to address large-scale problems of practical interest.

## 4.7   Nomenclature

**Indices/Sets**

$i \in I$          Products
$f, f' \in F$     Product families (families)
$j \in J$         Processing units (units)
$n \in N$       Production periods (periods)

**Subsets**

$F_j$         Families that can be processed in unit $j$
$I_f$         Products that belong to family $f$
$I_f^{idle}$     Dummy product for family $f$ (one per family)
$I_j$         Products that can be processed in unit $j$
$J_f$        Processing units that can process family $f$
$J_i$         Processing units that can process product $i$

**Parameters**

| | |
|---|---|
| $\gamma_{ff'j}$ | Changeover time between family $f$ and $f'$ in unit $j$ |
| $\delta_{ij}$ | Setup time of product $i$ in unit $j$ |
| $\varsigma_{in}$ | Demand of product $i$ at time $n$ |
| $\theta_{ij}$ | Setup cost of product $i$ in unit $j$ |
| $\lambda_{ij}$ | Operating cost of product $i$ in unit $j$ |
| $\xi_{in}$ | Holding cost of product $i$ in period $n$ |
| $\rho_{ij}^{\max}$ | Maximum production rate of product $i$ in unit $j$ |
| $\rho_{ij}^{\min}$ | Minimum production rate of product $i$ in unit $j$ |
| $\tau_{ij}^{\max}$ | Maximum processing time of product $i$ in unit $j$ |
| $\tau_{ij}^{\min}$ | Minimum processing time of product $i$ in unit $j$ |
| $\phi_{ff'j}$ | Changeover cost between family $f$ and $f0$ in unit $j$ |
| $\psi_{in}$ | Backlog cost of product $i$ in period $n$ |
| $\omega_{jn}$ | Available production time in unit $j$ in period $n$ |

**Continuous Variables**

| | |
|---|---|
| $B_{in}$ | Backlog of product $i$ at time $n$ |
| $C_{fjn}$ | Completion time for family $f$ in unit $j$ in period $n$ |
| $P_{in}$ | Total produced amount of product $i$ in period $n$ |
| $Q_{ijn}$ | Produced amount of product $i$ in unit $j$ during period $n$ |
| $S_{in}$ | Inventory of product $i$ at time $n$ |
| $T_{ijn}$ | Processing time for product $i$ in unit $j$ in period $n$ |
| $T_{fjn}^{F}$ | Processing time for family $f$ in unit $j$ in period $n$ |
| $U_{jn}$ | Time within period $n$ consumed by a changeover operation that will be $\bar{U}_{jn}$ time within period $n$ consumed by a changeover operation that started in the next period on unit $j$ |

**Binary Variables**

| | |
|---|---|
| $Wf_{fjn}$ | $= 1$, If family $f$ is assigned first to unit $j$ in period $n$ |
| $WL_{fjn}$ | $= 1$, If family $f$ is assigned last to unit $j$ in period $n$ |
| $X_{ff'jn}$ | $= 1$, If family $f$ is processed exactly before $f0$ in period $n$ in unit $j$ |
| $\bar{X}_{ff'jn}$ | $= 1$, If family $f$ in period $n-1$ is followed from family $f'$ in period $n$ on unit $j$ |
| $Y_{ijn}$ | $= 1$, If product $i$ is assigned to unit $j$ in period $n$ |
| $Y_{jn}^{F}$ | $= 1$, If family $f$ is assigned to unit $j$ in period $n$ |

# References

Grunow M, Günther HO, Lehmann M (2002) Campaign planning for multi-stage batch processes in the chemical industry. OR Spectrum 24:281–314

Günther HO, Grunow M, Neuhaus U (2006) Realizing block planning concepts in make-and pack production using MILP modelling and SAP APO. Int J Prod Res 44(18–19):3711–3726

Inman RR, Jones PC (1993) Decomposition for scheduling flexible manufacturing systems. Oper Res 41(3):608–617

Karmarkar US, Schrage L (1985) The deterministic dynamic product cycling problem. Oper Res 33(2):326–345

Kopanos GM, Puigjaner L, Georgiadis MC (2010) Optimal production scheduling and lotsizing in dairy plants: the yogurt production line. Ind Eng Chem Res 49(2):701–718

Kopanos GM, Puigjaner L, Maravelias CT (2011) Production planning and scheduling of parallel continuous processes with product family considerations. Ind Eng Chem Res 50:1369–1378

Sahinidis NV, Grossmann IE (1991) MINLP model for cyclic multiproduct scheduling on continuous parallel lines. Comput Chem Eng 15:85–103

Sung C, Maravelias CT (2008) A mixed-integer programming formulation for the general capacitated lot-sizing problem. Comput Chem Eng 32(1–2):244–259

# Part III
# Semicontinuous Processes

# Chapter 5
# Production Scheduling in Multistage Semicontinuous Process Industries

## 5.1 Introduction

Production scheduling is certainly important in the process industries including production of specialty chemicals, pharmaceuticals, food, and paper in which material is often produced in campaigns using various batch sizes in shared equipments. Determining the timing and sequence of production campaigns becomes increasingly difficult as manufacturers strive to achieve increased production rates while minimizing total costs. In this chapter, a real-life multiproduct multistage ice-cream production facility is considered as a representative semicontinuous process industry. Most production plants in the food industry sector combine continuous operations and batch processes in their product processing routes, thus working in semicontinuous production mode, since production is more flexible and equipment can be more efficiently utilized.

A Mixed Integer Programming (MIP) framework and a solution strategy are presented for the optimal production scheduling of multiproduct multistage semicontinuous process industries, such as the ice-cream production facility studied in details (Kopanos et al. 2011). The overall mathematical framework relies on an efficient modeling approach of the sequencing decisions, the integrated modeling of all production stages, and the inclusion of strong valid integer cuts in the formulation. The simultaneous optimization of all processing stages increases the plant production capacity, reduces the production cost for final products, and facilitates the interaction among the different departments of the production facility. To the best of our knowledge, there is no previous work in the literature presenting an exact method for addressing the challenges of the underlying food process scheduling problem.

© Springer Nature Switzerland AG 2019

G. M. Kopanos and L. Puigjaner, *Solving Large-Scale Production Scheduling and Planning in the Process Industries*,
https://doi.org/10.1007/978-3-030-01183-3_5

## 5.2   The Ice-Cream Production Facility

The ice-cream production facility under study—that represents a typical ice-cream factory—was first described by Bongers and Bakker (2006). The plant is based on a three-stage production process, as shown in Fig. 5.1. More specifically, the basic mixes are produced in the main process line (PROC), followed by storage in the aging vessels (V1–V6). After a minimum aging time, the mixes are used for the production of final products (A–H) in two packing lines (PACK1, PACK2). The main process line has a processing rate of 4.5 tons/h and can feed all aging vessels; one at a time. Packing line 1 is supplied by two aging vessels (V1 and V2) of 8 tons capacity each and can accommodate products A–D, whereas packing line 2 can pack products E–H and it is supplied by four aging vessels (V3–V6) of 4 tons capacity each. Figure 5.1 shows the minimum aging times and packing rates for all products. The maximum shelf life for all intermediate mixes in aging stage is 72 h. Sequence-dependent changeover, or simply changeover, operations (mainly cleaning and sterilizing tasks) are performed both in the process and the packing lines whenever the production is changed from a product to a different one. Table 5.1 provides the necessary changeover times for performing these operations in the process and the packing lines. Moreover, a cleaning time of 2 h is needed before

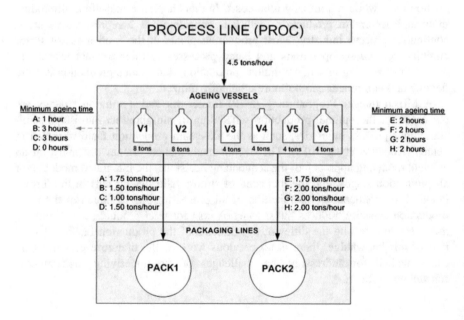

**Fig. 5.1**  Ice-cream production facility

**Table 5.1** Changeover times in the process line and the packing lines (minutes)

| Product | A | B | C | Process D | Line E | F | G | H | A | B | Packing C | Lines D | E | F | G | H |
|---|---|---|---|---|---|---|---|---|---|---|---|---|---|---|---|---|
| A | 0 | 30 | 30 | 30 | 30 | 30 | 30 | 30 | 0 | 60 | 60 | 60 | 0 | 0 | 0 | 0 |
| B | 30 | 0 | 30 | 30 | 30 | 30 | 30 | 30 | 30 | 0 | 60 | 60 | 0 | 0 | 0 | 0 |
| C | 30 | 30 | 0 | 30 | 30 | 30 | 30 | 30 | 30 | 30 | 0 | 60 | 0 | 0 | 0 | 0 |
| D | 30 | 30 | 30 | 0 | 30 | 30 | 30 | 30 | 30 | 30 | 30 | 0 | 0 | 0 | 0 | 0 |
| E | 30 | 30 | 30 | 30 | 0 | 15 | 15 | 15 | 0 | 0 | 0 | 0 | 0 | 60 | 60 | 60 |
| F | 30 | 30 | 30 | 30 | 5 | 0 | 15 | 15 | 0 | 0 | 0 | 0 | 30 | 0 | 60 | 60 |
| G | 30 | 30 | 30 | 30 | 5 | 5 | 0 | 15 | 0 | 0 | 0 | 0 | 30 | 30 | 0 | 60 |
| H | 30 | 30 | 30 | 30 | 5 | 5 | 5 | 0 | 0 | 0 | 0 | 0 | 30 | 30 | 30 | 0 |

shutting down the process line and the packing machines. Finally, the production facility is available for 120 h a week (a 48-h weekend).

## 5.3 Typical Scheduling Practice in Food Industries

As Bongers and Bakker (2006) pointed out, the practical scheduling inside the vast majority of food factories is focused on just scheduling the packing lines. Afterward, the packing lines schedule is "thrown over the wall" to the process department, in which a schedule should be made to meet the packing demand. To go further, this schedule is also "thrown over the wall" to the incoming materials department, in which a schedule is made to order/receive the materials. The way food industries are being scheduled nowadays is posing two major problems:

- Each department will strive to ensure that is not to blame for not delivering packing products according to the schedule, while less available production capacity will be communicated to the plant management.
- Any change in the packing schedule might lead to an infeasible schedule in the upstream departments. For instance, packing lines may not run due to lack of intermediate products or unnecessary intermediate products being made in the process plant floor.

Since the above problems are frequently met in relevant industrial environments, the challenge is to appropriately tackle them in an integrated way in order to increase the plant production capacity and reduce the production cost for final products. The gap between scheduling theory and practice is still evident, since most academic developments are too distant from industrial environments. This industrial reality drove Bongers and Bakker (2006) to characterize as a challenging problem the simultaneous scheduling of all processing stages (i.e., the process line, the aging vessels, and the packing lines) in typical food processing industries, such as the ice-cream production facility studied in details.

## 5.4  Problem Statement

This study considers the production scheduling problem of industrial-scale multi-product multistage semicontinuous processes, similar to the previously described ice-cream production process, with the following features:

(i)   A set of product orders $i \in I$ should be processed by following a predefined sequence of processing stages $s \in S$ with processing units $j \in J$ working in parallel.

(ii)  The total demand $\varsigma_i$ for each product order $i$ is divided into a number of batches $b \in B$ that must follow a specific set of processing stages $s$.

(iii) Product order $i$ can be processed in a specific subset of units $j \in J_i$. Similarly, processing stage $s$ can be processed in a specific subset of units $j \in J_s$.

(iv)  Every aging vessel $j \in J_{s_2}$ has a maximum capacity $\mu_j^{\max}$. In aging vessels, a product batch should remain for a minimum aging time $\tau_i^{age} i$ and no longer than its corresponding shelf life $\varepsilon_i^{life}$.

(v)   Parameter $\rho_{ij}$ denotes the processing and packing rate for every product $i$ in unit $j \in J_i$.

(vi)  Sequence-dependent changeover times $\gamma_{ii'j}$ between consecutive product orders are present in the process (S1) and the packing (S3) stage.

(vii) All model parameters are deterministic.

(viii) Once the processing of an order in a given stage is started, it should be carried out until completion without interruption (i.e., non-preemptive mode).

The key decision variables are as follows:

(i)   the allocation of batch $b$ of product $i$ to units $j \in J_i$ per stage, $Y_{ibs\,j}$;

(ii)  the relative sequence for any pair of product batches $i$, $b$ and $i'$, $b'$ in the process line (i.e., stage S1), $\overline{X}_{ibi'b'}$;

(iii) the relative sequence for any pair of products $i$ and $i'$ in aging vessels (i.e., stage S2) and packing lines (i.e., stage S3) for $j \in J_i \cap J_{i'} \cap J_s$, $X_{ii'}$; and

(iv)  the starting and completion time for batch $b$ of product $i$ in stage $s$; $L_{ibs}$ and $C_{ibs}$, respectively.

The minimization of makespan constitutes the optimization goal in this case study.

### 5.4.1  Industrial Production Policy

Generally speaking, in most food processing industries, such as the one studied here, in order to achieve high production levels and minimize switchovers of products, the industrial practice imposes operations of the intermediate storage/processing vessels (e.g., aging vessels, fermentation tanks, etc.) in their maximum capacity (Kopanos et al. 2012). Coming back to the ice-cream production facility under study, the fact that the aging vessels operate in maximum capacity allows us to solve the batching problem beforehand, and afterward solve the scheduling problem for the predefined number of product batches.

**Minimum Number of Batches**  The minimum number of batches $\beta_i^{\min}$ to satisfy the demand for each specific product order $i$ depends on the capacity of the storing/aging vessels in which it can be stored. In the case that these aging vessels have the same capacity $\mu_j^{\max}$, the minimum number of batches is given by

$$\beta_i^{\min} = \frac{\varsigma_i}{\mu_j^{\max}} \quad \text{where } j \in (J_i \cap J_{S_2})$$

**Filling Time for Aging Vessels (Processing Time in the Process Line)**  The time, $\tau_i^{\text{fill}}$, to fill an aging vessel with a product $i$ is calculated by

$$\tau_i^{\text{fill}} = \frac{\mu_j^{\max}}{\rho_{ij'}} \quad \text{where } j \in (J_i \cap J_{S_2}), j' \in (J_i \cap J_{S_1})$$

Note that unit $j' \in J_{S_1}$ corresponds to the process line and unit $j \in J_{S_2}$ to the aging vessels. It should be also noticed that the above equation is valid *if and only if* (i) the product $i$ can be stored into a number of *equal-capacity* aging vessels $\mu_j^{\max}$, and (ii) the aging vessels are supplied by the process lines that have the same processing rate $\rho_{ij'}$. Obviously, the aging vessels' filling time equals to the processing time in the process line due to the continuous process mode.

**Emptying Time for Aging Vessels (Packing Time in the Packing Lines)**  The time $\tau_i^{\text{empty}}$ to empty an aging vessel from a product $i$ is calculated as follows:

$$\tau_i^{\text{empt}} = \frac{\mu_j^{\max}}{\rho_{ij'}} \quad \text{where } j \in (J_i \cap J_{S_2}), j' \in (J_i \cap J_{S_1}), j' \in (J_i \cap J_{S_3})$$

Note that unit $j' \in (J_i \cap J_{S_3})$ corresponds to the packing line where product $i$ can be packed. Once again, this expression is valid *if and only if* the product $i$ can be stored into a number of *equal-capacity* aging vessels $\mu_j^{\max}$. Obviously, the aging vessels' emptying time is equal to the packing time of the packing lines.

## 5.5  Mathematical Formulation

In this section, the proposed MIP formulation is presented for the production scheduling of the multiproduct multistage production facility described above. Constraints are grouped according to the type of decision (e.g., assignment, timing, sequencing, etc.). To facilitate the presentation of the model, we use uppercase

Latin letters for optimization variables and sets, and lowercase Greek letters for parameters.

**Unit Allocation Constraints for Any Product Batch in Every Processing Stage**
Constraints (5.1) guarantee that each product batch $i$, $b$ goes through one unit $J \in (J_i \cap J_S)$ in each stage $s$.

$$\sum_{J \in (J_i \cap J_S)} Y_{ibsj} = 1 \quad \forall i, b \leq \beta_i^{\min}, s \tag{5.1}$$

**Timing Constraints for a Product Batch in the Same Processing Stage** The timing for a batch $b$ of product $i$ in each stage $s$ is defined by constraint sets (5.2)–(5.5) (see Fig. 5.2). In the process stage, the completion time $C_{ibs}$ for a batch $b$ of product $i$ equals to its starting time $L_{ibs}$ plus the necessary aging vessel filling time $\tau_i^{fill}$, according to constraints (5.2). In the aging stage, the timing for each product batch $i$, $b$ is given by constraints (5.3). The standing (waiting) time for a product batch $i$, $b$ in aging stage is denoted by $W_{ibs}$. This standing time plus the minimum aging time $\tau_i^{age}$ should not exceed the product shelf life, as constraint set (5.4) ensures. Finally, constraints (5.5) calculate the timing for any batch $b$ of product $i$ in the packing stage.

$$L_{ibs} + \tau_i^{fill} = C_{ibs} \quad \forall i, b \leq \beta_i^{\min}, s = 1 \tag{5.2}$$

$$L_{ibs} + \tau_i^{fill} + \tau_i^{age} + W_{ibs} + \tau_i^{empt} = C_{ibs} \quad \forall i, b \leq \beta_i^{\min}, s = 2 \tag{5.3}$$

$$W_{ibs} \leq \varepsilon_i^{life} - \tau_i^{age} \quad \forall i, b \leq \beta_i^{\min}, s = 2 \tag{5.4}$$

$$L_{ibs} + \tau_i^{empt} = C_{ibs} \quad \forall i, b \leq \beta_i^{\min}, s = 3 \tag{5.5}$$

**Timing Constraints for a Product Batch Between Consecutive Processing Stages** Constraints (5.6) and (5.7) define the timing for every product batch $i$, $b$ between two consecutive processing stages (see Fig. 5.2). Constraints (5.6) state that the starting time $L_{ibs}$, for any product batch $i$, $b$, in the aging stage is equal to the starting time in the process stage due to the continuous nature of the process stage.

Moreover, an aging vessel is free for processing a product only when it is completely empty; therefore, the completion time of a product batch $i$, $b$ stored in an aging vessel equals the completion time of this batch in the packing line, according to constraints (5.7).

$$L_{ibs} = L_{ibs-1} \quad \forall i, b \leq \beta_i^{\min}, s = 2 \tag{5.6}$$

$$C_{ibs} = C_{ibs-1} \quad \forall i, b \leq \beta_i^{\min}, s = 3 \tag{5.7}$$

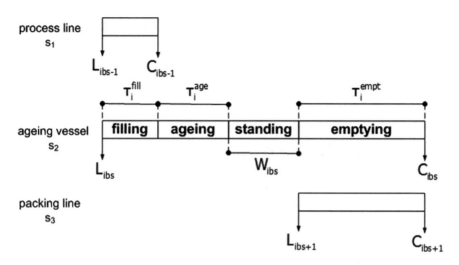

**Fig. 5.2** Timing decisions for a product batch $i$, $b$ for every processing stage

**Timing Constraints for Two Batches of the Same Product in the Packing Stage** If the underlying industrial policy requires a single production campaign in the packing stage, without allowing a waiting time between batches of the same product, constraints (5.8) must be added into the MIP formulation. According to these constraints, the completion time for a product batch $i$, $b$ should be equal to the starting time for the next indexed product batch $i$, $b + 1$.

$$C_{ibs} = L_{ibs+1s} \quad \forall i, b \leq \beta_i^{min}, s = 3 \tag{5.8}$$

**Sequencing Constraints Between Product Batches in all Processing Stages** Constraints (5.9)–(5.13) define the relative sequencing between two product batches. Constraints (5.9)–(5.12) have been formulated as big-M constraints, where the available scheduling horizon $\omega$ plays the role of the M parameter. In addition, our mathematical formulation uses global precedence sequencing variables: (i) for any pair of product batches $i$, $b$ and $i'$, b' $(i < i')$ in the process stage $\bar{X}_{ibi'b'}$, and (ii) for any pair of different products $i$ and $i'$ $(i < i')$ both in the aging and the packing stage $X_{ii'}$. Note that the sequencing decisions are the same for the aging and packing stage, and they are modeled for both stages through a single binary variable $X_{ii'}$ for a given pair of products. Constraints (5.9) enforce the starting time of a product batch $i'$, b' to be greater than the completion time of any product batch $i$, $b$ processed beforehand plus the corresponding sequence-dependent changeover time $\gamma_{ii'j}$, when both batches are assigned to the same process line. Similarly, constraint set (5.10) describes the opposite case. In a similar manner, constraints (5.11) and (5.12) define the sequencing between any pair of different products $i$ and $i' > i$ in the aging and the packing stage. Finally, in order to avoid symmetric solutions, if two batches $b$ and $b' > b$ of the same product $i$ are assigned to the same unit, we assume that the lower indexed batch $b$ is performed first, according to constraints (5.13).

$$L_{i'b's} \geq C_{ibs} + \gamma_{ii'j} - \omega\left(1 - \bar{X}_{ibi'b'}\right) - \omega\left(2 - Y_{ibsj} - Y_{ib'sj}\right)$$
$$\forall i, b \leq \beta_i^{\min}, i', b' \leq \beta_{i'}^{\min}, s, j \in (J_i \cap J_{i'} \cap J_s) : i < i', s = 1 \tag{5.9}$$

$$L_{ibs} \geq C_{i'b's} + \gamma_{i'ij} - \omega\bar{X}_{ibi'b'} - \omega\left(2 - Y_{ibsj} - Y_{ib'sj}\right)$$
$$\forall i, b \leq \beta_i^{\min}, i', b' \leq \beta_{i'}^{\min}, s, j \in (J_i \cap J_{i'} \cap J_s) : i < i', s = 1 \tag{5.10}$$

$$L_{ibs} \geq C_{i'b's} + \gamma_{i'ij} - \omega\bar{X}_{ibi'b'} - \omega\left(2 - Y_{ibsj} - Y_{ib'sj}\right)$$
$$\forall i, b \leq \beta_i^{\min}, i', b' \leq \beta_{i'}^{\min}, s, j \in (J_i \cap J_{i'} \cap J_s) : i < i', s > 1 \tag{5.11}$$

$$L_{i,b,s} \geq C_{i'b's} + \gamma_{i'ij} - \omega X_{ii'} - \omega\left(2 - Y_{ibsj} - Y_{ib'sj}\right)$$
$$\forall i, b \leq \beta_i^{\min}, i', b' \leq \beta_{i'}^{\min}, s, j \in (J_i \cap J_{i'} \cap J_s) : i < i', s > 1 \tag{5.12}$$

$$L_{i,b',s} \geq C_{ibs} - \omega\left(2 - Y_{ibsj} - Y_{ib'sj}\right)$$
$$\forall i, b \leq \beta_i^{\min}, b' \leq \beta_i^{\min}, s, j \in (J_i \cap J_s) : b < b' \tag{5.13}$$

**Objective Function: Makespan** The time point at which all product orders are accomplished corresponds to the makespan. The makespan objective is closely related to the throughput objective. For instance, minimizing the makespan in a parallel-machine environment with changeover times forces the scheduler to balance the load over the various machines and to minimize the sum of all the setup times in the critical bottleneck path. Moreover, the minimization of makespan probably leads to a maximization of production at a mid-term planning level:

$$\min \quad C_{\max} \geq C_{ibs} \quad \forall i, b \leq \beta_i^{\min}, s = 3 \tag{5.14}$$

**Tightening Constraints** To reduce the computational effort, constraints (5.15) can further tighten the mathematical formulation by imposing a lower bound on the makespan objective. Note that parameter $\phi_j^{\min}$ represents the minimum waiting time to begin using packing line $j \in J_{s_3}$. Obviously, $\phi_j^{\min}$ depends on the minimum filling time for aging vessels $j' \in J_{s_2}$ that are connected to packing line $j$. Additionally, parameter $\alpha_j^{\min}$ stands for the minimum number of products that should be assigned to packing line $j$ to ensure full demand satisfaction. Finally, parameter $\gamma_j^{\min}$ denotes the minimum changeover time between two different products in packing line $j$.

$$C_{\max} \geq \phi_j^{\min} + (\alpha_j^{\min} - 1)\gamma_j^{\min} + \sum_{i \in I_j} \tau_i^{\text{empt}} \beta_i^{\min}$$
$$\forall s, j_s : s = 3 \tag{5.15}$$

The proposed MIP model can be further tightened by correlating the relative sequence variables of the process stage (S1) and the packing stage (S3). Constraints

(5.16) describe these valid integer cuts by forcing the relative sequence between products $i$ and $i' > i$ in packing and aging stages to maintain the same for the product batches $i$, $b$ and $i'$, $b'$ in the process stage; for products $i$ and $i'$ that share the same packing line. In other words, if product $i$ is assigned before product $i'$ to packing unit $(J_i \cap J_{i'} \cap J_{s_3})$, constraint set (5.16) drives all the batches of product $i$ to be allocated to the process line before any batch of product $i'$. Figure 5.3 illustrates graphically the role of these constraints.

$$\bar{X}_{ibi'b'} = X_{ii'} \quad \forall i, b \le \beta_i^{\min}, i', b' \le \beta_{i'}^{\min}, s, j \in (J_i \cap J_{i'} \cap J_s),$$
$$j' \in (J_i \cap J_{i'} \cap J_{s+2}) : i < i', s = 1 \tag{5.16}$$

**Integrality and Nonnegativity Constraints** The domains of all decision variables are defined as follows:

$$Y_{ibsj} \in \{0, 1\} \, \forall i, b \le \beta_i^{\min}, s, j \in (J_i \cap J_s)$$
$$\bar{X}_{ibi'b'} \in \{0, 1\} \quad \forall i, b \le \beta_i^{\min}, i', b' \le \beta_{i'}^{\min},$$
$$s, j \in (J_i \cap J_{i'} \cap J_s) : i < i', s = 1$$
$$X_{ii'} \in \{0, 1\} \quad \forall i, i', s, j \in (J_i \cap J_{i'} \cap J_s) : i < i', s > 2 \tag{5.17}$$
$$L_{ibs}, C_{ibs} \ge 0 \quad \forall i, b \le \beta_i^{\min}, s$$
$$W_{ibs} \ge 0 \quad \forall i, b \le \beta_i^{\min}, s = 2$$

The overall MIP formulation consists of constraint sets (5.1)–(5.17).

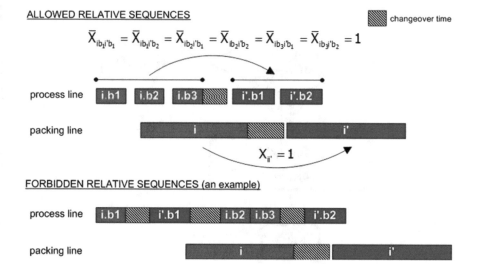

**Fig. 5.3** Illustrative example: relative sequences according to constraints (5.16)

## 5.6  Proposed Solution Methodology

In this section, a solution methodology is presented for solving efficiently the scheduling problem under study. Before explaining the proposed solution technique, the following points should be taken into consideration: (i) final products can be packed into a specific packing line, (ii) the intermediates of final products that are packed into the same packing line can be stored into the same equal-capacity aging vessels, and (iii) full demand satisfaction is imposed.

In accordance with the abovementioned points, a lower bound for the makespan for every packing line $\zeta_j^{min}$ is calculated as follows:

$$\zeta_j^{min} = \phi_j^{min} + \gamma_j^{total} + \sum_{i \in I_j} \tau_i^{empt} \beta_i^{min} \quad \forall s, j \in J_s : s = 3$$

where parameter $\gamma_j^{total}$ represents the minimum total changeover time in packing line $j$. Additionally, we define the subset $J_{min}$ which contains the packing line that appears the highest $\zeta_{min}$ value. By doing this, it can be safely assumed that the makespan $C_{max}$ is equal to the makespan of the packing line $J \in J_{min}$, as follows:

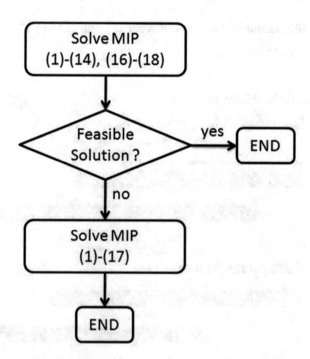

**Fig. 5.4**  Proposed solution methodology

$$C_{\max} = \phi_j^{\min} + \gamma_j^{\text{total}} + \sum_{i \in I_j} \tau_i^{\text{empt}} \beta_i^{\min} \quad \forall s, j \in J_s : s = 3 \qquad (5.18)$$

Therefore, constraints (5.15) can be placed by constraints (5.18), which are tighter. Note that constraints (5.18) provide the minimum possible makespan, and their incorporation into the MIP model may violate some of the remaining constraints, thus leading to infeasible solutions. This may happen when the packing line $J \in J^{\min}$ is not the bottleneck, and the timing decisions depend on the previous processing stages. In this case, only constraints (5.15) should be used.

As Fig. 5.4 demonstrates, the proposed solution methodology can be distinguished into two steps:

1. Solve the MIP model consisting of constraints (5.1)–(5.14), and (5.16)–(5.18). The solution method terminates, if a feasible solution is obtained. Otherwise, go to step 2.
2. Solve the MIP formulation consisting of constraints (5.1)–(5.17).

## 5.7  Industrial Case Study

A real-life industrial case study, as described in Sect. 5.2, is used to illustrate the applicability and the efficiency of the proposed scheduling approach and solution strategy. A total set of ten different problem instances, regarding the demands for final products, has been solved. Roughly speaking, final products are characterized by very high demands given in Table 5.2. All problem instances have been solved in a Dell Inspiron 1520 2.0 GHz with 2 GB RAM using CPLEX 11 via a GAMS 22.8 interface.

**Problem Instance PI.01** Bongers and Bakker (2006) made the first attempt to solve this scheduling problem using advanced commercial scheduling software. As they have reported, a feasible schedule on all stages could not be derived automatically by applying the available solvers. They finally obtained a feasible schedule by manual interventions. Recently, Subbiah and Engell (2010) studied the same ice-cream production plant. They used the framework of timed automata, and they solved the optimization problem using reachability analysis (Abdeddaïm et al. 2006). A heuristic methodology was implemented to reduce the model size. A feasible solution was found in few CPUs; however, it cannot be ruled out that the heuristics employed pruned the optimal solution.

In this problem instance, some decisions in the beginning of the production week of interest have been taken at the end of the previous production week. More specifically, product batches D.b1, G.b1, G.b2, and G.b3 have already passed from the process line and assigned to aging vessels V1, V3, V4, and V5, respectively. Moreover, these product batches have already allocated to the aging process at the

**Table 5.2** Demands for final products for all problem instances (kg)

| Products | PL01 | PL02 | PL03 | PL04 | PL05 | PL06 | PL07 | PL08 | PL09 | PL10 |
|---|---|---|---|---|---|---|---|---|---|---|
| A | 80,000 | 48,000 | 32,000 | 8000 | 88,000 | 16,000 | 8000 | 16,000 | 48,000 | 8000 |
| B | 48,000 | 56,000 | 32,000 | 32,000 | 16,000 | 16,000 | 8000 | 40,000 | 24,000 | 72,000 |
| C | 32,000 | 16,000 | 40,000 | 64,000 | 24,000 | 16,000 | 96,000 | 32,000 | 56,000 | 8000 |
| D | 8000 | 45,000 | 32,000 | 24,000 | 40,000 | 88,000 | 8000 | 56,000 | 16,000 | 72,000 |
| E | 112,000 | 80,000 | 32,000 | 52,000 | 12,000 | 24,000 | 116,000 | 36,000 | 8000 | 80,000 |
| F | 12,000 | 44,000 | 60,000 | 44,000 | 48,000 | 24,000 | 64,000 | 40,000 | 92,000 | 80,000 |
| G | 48,000 | 12,000 | 44,000 | 88,000 | 64,000 | 104,000 | 4000 | 60,000 | 20,000 | 4000 |
| H | 24,000 | 64,000 | 80,000 | 32,000 | 84,000 | 52,000 | 4000 | 60,000 | 88,000 | 32,000 |

beginning of the time horizon ($t = 0$), and as such they are ready for passing to the packing stage again at $t = 0$. For this reason, in this example, parameter $\phi_j^{\min} = 0$.

The MIP model consists of 15,848 equations, 491 continuous variables, and 2024 binary variables. The optimal solution was reached in just 1.83 CPUs despite the fact of the challenging (very high) total demand for final products. The optimal production schedule, which is illustrated in Fig. 5.5, results in a makespan of 118.55 h. Table 5.3 shows the breakdown of the utilization of the available scheduling time in the process and packing lines. The process line is utilized for both processing and cleaning 80.89% of the available time compared to a food industry standard of 70%. Packing lines 1 and 2 operate at 98.79 and 92.08% of the total available time, respectively, including both packing and cleaning. The high total demand explains the high utilization in the process and the packing lines. Packing lines illustrate low total changeover times. As expected, in the process line, total cleaning times are higher, since changeovers for batches of different products are more frequent. Finally, it is worth mentioning that in the feasible schedule reported by Bongers and Bakker (2006), the process line is utilized 90% (that is 9.11% higher than that of the optimal schedule of this work) of the available time, thus resulting in higher production costs (due to higher changeover costs) comparing it with the proposed optimal production schedule.

In general, it should be mentioned that solution strategies that do not optimally integrate the scheduling of all processing stages (i.e., process line, aging vessels, and packing lines) face the risk of not generating optimal solutions. In this specific (high demand) case study, these solution strategies probably cannot give a feasible schedule. In other words, they may propose solutions where full demand

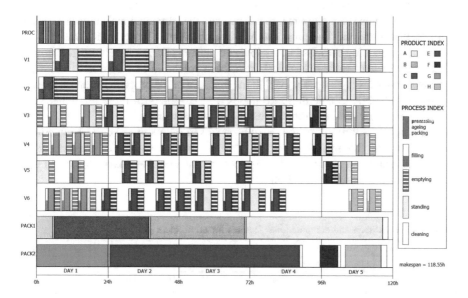

**Fig. 5.5** Optimal production schedule for PI.01 (minimization of makespan)

**Table 5.3** Process line and packing lines utilization breakdown for PI.01

| Processing unit | Unit operation | Time (h) | Operation utilization (%) | Total unit utilization (%) |
|---|---|---|---|---|
| PROC | Processing | 76.48 | 67.73 | 80.09 |
| | Cleaning | 20.58 | 17.15 | |
| | Idle | 22.94 | 19.11 | |
| PACK1 | Packing | 115.05 | 95.88 | 98.79 |
| | Cleaning | 3.50 | 2.92 | |
| | Idle | 1.45 | 1.21 | |
| PACK2 | Packing | 106.00 | 88.33 | 92.08 |
| | Cleaning | 4.50 | 3.75 | |
| | Idle | 9.50 | 7.92 | |

satisfaction is not achieved inside the available production horizon. Manual intervention may still be necessary in order to obtain feasible (i.e., full demand satisfaction), and probably not optimal schedules (Bongers and Bakker 2006).

**Problem Instances PI.02–PI.10** In problem instances PI.02–PI.10, we consider no overlapping decisions from the previous week schedule. This fact allows us to predefine the relative sequence for products in each packing line, taking into account the sequence-dependent changeover times included in Table 5.1. It can be observed that the optimal relative sequence with respect to the minimization of changeover times in PACK1 is $D \rightarrow C \rightarrow B \rightarrow A$, and in PACK2 is $H \rightarrow G \rightarrow F \rightarrow E$. That means that $X_{iD} = X_{BC} = X_{AC} = X_{AB} = 0$ and $X_{iD} = X_{BC} = X_{AC} = X_{AB} = 0$ in PACK1 and PACK2, respectively.

Table 5.4 presents the optimal makespan and the computational characteristics for all problem instances. The proposed MIP formulation in tandem with the proposed solution methodology results in very low computational times for all cases. It is noted that seven of ten problem instances have been solved in less than 2 CPUs. These problem instances have been solved in the first step of the proposed solution method.

It is worthwhile to note that zero nodes were explored for these problems. Also, notice that the remaining problem instances, which were infeasible in the first step of the proposed solution method, have been solved in less than a CPU minute in the second step of the solution method. Despite the complexity of the scheduling problems addressed in this work, all problem instances have been solved to optimality with low computational effort.

**Table 5.4** Makespan and computational features for all problem instances

| | PL01 | PL02 | PL03 | PL04 | PL05 | PL06 | PL07 | PL08 | PL09 | PL10 |
|---|---|---|---|---|---|---|---|---|---|---|
| $C_{max}$ | 118.55 | 118.04 | 116.67 | 118.10 | 116.90 | 110.10 | 116.52 | 110.42 | 115.37 | 113.85 |
| CPUs | 1.83 | 0.88 | 23.20 | 51.41 | 1.22 | 15.70 | 0.39 | 0.58 | 0.54 | 0.75 |
| Nodes | 0 | 0 | 467 | 751 | 0 | 510 | 0 | 0 | 0 | 0 |

[a]Makespan includes 2 h of cleaning before shutting down the packing lines

## 5.8 Concluding Remarks

In this chapter, a mathematical programming framework and an efficient solution approach have been proposed for the production scheduling in food process industries similar to the ice-cream production facility studied in details. This model can easily be the core element of a computer-aided advanced scheduling and planning system in order to facilitate decision-making in relevant industrial environments. As the challenging case study reveals, the proposed approach features a salient computational performance due to the efficient modeling approach of the sequencing decisions, and the strong valid integer cuts introduced. However, it should be mentioned that in extremely large-scale scheduling problems, potentially involving hundreds of products, the proposed MIP model might result in huge model sizes difficult to be solved within a reasonable (acceptable) computational time. In that case, the proposed mathematical formulation can be easily used as the core MIP model in the MIP-based decomposition strategy described in Chap. 8, in an attempt to make it attractive for the solution of complex large-scale industrial scheduling problems. Finally, it is worth noting that the proposed MIP model is well-suited to a real-life ice-cream production facility; however, it could be also used, with minor modifications, in scheduling problems arising in other semicontinuous industries with similar processing features (e.g., yogurt production lines, milk processing plants, etc.).

## 5.9 Nomenclature

### Indices/Sets

| | |
|---|---|
| $b, b' \in B$ | Product batches (batches) |
| $i, i' \in I$ | Product orders (products) |
| $j, j' \in J$ | Processing units (units) |
| $s \in S$ | Processing stages (stages) |

### Subsets

| | |
|---|---|
| $I_j$ | Products $i$ that can be processed in unit $j$ |
| $J_i$ | Available units $j$ to process product $i$ |
| $J_s$ | Available units $j$ to process stage $s$ |
| $J^{min}$ | Packing line that appears the highest lower bound for unit makespan |

### Parameters

| | |
|---|---|
| $\alpha_j^{min}$ | Minimum number of products assigned to packing line $j \in J_{s_3}$ |
| $\beta_i^{min}$ | Minimum number of batches for product $i$ |
| $\gamma_{ii'j}$ | Sequence-dependent changeover time between orders $i$ and $i0$ in unit $j \in (J_i \cap J_{i'})$ |
| $\gamma_j^{min}$ | |

|  | Minimum sequence-dependent changeover time between two different products in packing line $j \in J_{s_3}$ |
| --- | --- |
| $\gamma_j^{\text{total}}$ | Minimum total sequence-dependent changeover time in packing line $j \in J_{s_3}$ |
| $\varepsilon_i^{\text{life}}$ | Shelf life for product $i$ in aging vessels |
| $\zeta_i$ | Demand for product $i$ |
| $\mu_j^{\text{max}}$ | Maximum capacity of aging vessel $j \in J_{s_2}$ |
| $\zeta_j^{\text{min}}$ | Lower bound for the makespan for packing line $j \in J_{s_3}$ |
| $\rho_{ij}$ | Processing rate for every product $i$ in the process line $j \in (J_i \cap J_{s_1})$ and the packing lines $j \in (J_i \cap J_{s_3})$ |
| $\tau_i^{\text{age}}$ | Minimum aging time for product $i$ |
| $\tau_i^{\text{empt}}$ | Emptying time of aging vessel for product $i$ |
| $\tau_i^{\text{fill}}$ | Filling time of aging vessel for product $i$ |
| $\phi_j^{\text{min}}$ | Minimum wait time to begin using packing line $j$ |
| $\omega$ | Available scheduling horizon |

**Continuous Variables**

$C_{ibs}$    Completion time for stage $s$ of batch $b$ of product $i$

$C_{\text{max}}$    Makespan

$L_{ibs}$    Starting time for stage $s$ of batch $b$ of product $i$

$W_{ibs}$    Standing (waiting) time for stage $s$ of batch $b$ of product $i$

**Binary Variables**

$X_{ibi'b'}$    = 1, If product $i$ is processed before product $i$ (for the aging vessels and the packing lines)

$\bar{X}_{ibi'b'}$    = 1, If batch $b$ of product $i$ is processed before batch $b$ of product $i$ (for the process line)

$Y_{ibsj}$    = 1, If stage $s$ of batch $b$ of product $i$ is assigned to unit $j$

# References

Abdedaim Y, Asarin E, Maler O (2006) Scheduling with timed automata. Theor Comput Sci 354:272–300

Bongers PMM, Bakker BH (2006) Application of multi-stage scheduling. In: Marquardt W, Pantelides C (eds) 16th European symposium on computer aided process engineering and 9th international symposium on process systems engineering, vol 21. Computer aided chemical engineering. Elsevier, Amsterdam, pp 1917–1922

Kopanos GM, Puigjaner L, Georgiadis MC (2011) Production scheduling in multiproduct multistage semicontinuous food processes. Ind Eng Chem Res 50(10):6316–6324

Kopanos GM, Puigjaner L, Georgiadis MC (2012) Efficient mathematical frameworks for detailed production scheduling in food processing industries. Comput Chem Eng 42((SI)):206–216

Subbiah S, Engell S (2010) Short-term scheduling of multi-product batch plants with sequence-dependent changeovers using timed automata models. In: Pierucci S, Buzzi-Ferraris G (eds) 20th European symposium on computer aided process engineering. Computer aided chemical engineering, vol 28. Elsevier, pp 1201–1206

# Chapter 6
# Resource-Constrained Production Planning and Scheduling in Multistage Semicontinuous Process Industries

## 6.1 Introduction

The production planning and scheduling problem in a single production site is usually concerned with meeting fairly specific production requirements. Customer orders, stock imperatives, or higher level supply chain or long-term planning would usually set these. Production planning and scheduling deals with the allocation over time of scarce resources between competing activities to meet these requirements in an efficient fashion. The key components of the resulting resource-constrained planning problem are resources, tasks, and time. The resources need not be limited to processing equipment items, but may include material storage equipment, transportation equipment (intra- and interplant), operators, utilities (e.g., steam, electricity, cooling water), auxiliary devices, and so on. The tasks typically comprise processing operations (e.g., reaction, separation, blending, and packing) as well as other activities that change the nature of materials and other resources such as transportation, quality control, cleaning, changeovers, etc. There are both external and internal elements of the time component. The external element arises out of the need to coordinate manufacturing and inventory with expected product listings or demands, as well as scheduled raw material receipts and even service outages. The internal element relates to executing the tasks in an appropriate sequence and at right times, taking into account of the external time events and resource availabilities. Overall, this arrangement of tasks over time and the assignment of appropriate resources to the tasks in a resource-constrained framework must be performed in an efficient fashion, which implies the optimization, as far as possible, of some objective. Typical objectives include the minimization of cost or the maximization of profit, the maximization of customer satisfaction, and the minimization of deviation from target performance (Shah 1998).

© Springer Nature Switzerland AG 2019
G. M. Kopanos and L. Puigjaner, *Solving Large-Scale Production Scheduling and Planning in the Process Industries*,
https://doi.org/10.1007/978-3-030-01183-3_6

## 6.2  Resource-Constrained Planning and Scheduling Aspects in Dairy Processing Industries

In the food processing industries, quite often, homogeneous products have to be packed (e.g., milk, yogurt). For an unpacked product, a variety of packing materials (cups, bins, etc) and packing sizes are available. Generally, the packing lines are used for various products in one type of packing material and various packing sizes. These possibilities make the scheduling of the packing lines rather complex (Van Dam et al. 1993). In recent years, under market pressure, the number of products in dairy plants has been increased, the order sizes have been reduced, and the delivery times have been shortened. This has caused augmented scheduling tasks and usually, the scheduling system supporting these tasks has not followed these changes sufficiently.

Of particular interest in the dairy industry is the production of yogurt. In order to make yogurt, milk has to be pasteurized and fermented to create batches of *white mass*, which are then placed in containers of various sizes, sometimes with fruit or other ingredients. A fresh dairy plant may make more than 10 types of white mass that differ, for example, in percentage of cream and type of yeast. A typical plant produces more than 100 final products on several packing lines corresponding to different flavors, sizes, and percentage of fat (Nakhla 1995). Between pasteurization, fermentation, and storage, as many as 100 tanks can be used. And because the plant produces food, extra care must be taken to ensure high standards of sanitation, control of allergens, batch traceability, and maximum product freshness. Figure 6.1 illustrates the main processing steps for producing stirred yogurt.

The plant must closely coordinate the two primary production steps: the transformation of raw materials—such as milk, milk powders, and yeast—into white

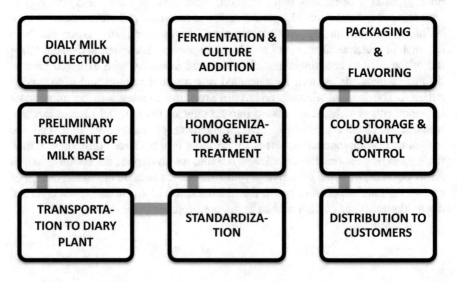

**Fig. 6.1**  Yogurt production process (except from set yogurt)

mass and the filling and packing of final products. The right amount of white mass has to be appropriately scheduled, and it has to be used as quickly as possible. Operational scheduling challenges include:

- Deciding which and how much white mass to produce in each tank given the available connections to the filling and packing lines.
- Finding the best time to clean the tanks and the filling lines given health and nutrition labeling requirements, and cleaning equipment availability.
- Synchronizing material consumption with white mass availability and freshness.
- Respecting batching policies for compliance with traceability regulations.
- Determining an optimal schedule for labor resources—maintaining a steady supply of finished goods within a minimum and maximum inventory corridor.
- Optimal production rescheduling under unexpected events such as modifications in product orders.

To this production, challenges must be added to those for high demand variability. Fresh dairy products are consumer goods with significant promotional marketing and a steady introduction of new products. Demand is often uncertain. New products may steal sales from old products or simply contribute to market share, and marketing campaigns can result in sales that are higher or lower than forecasted. And in the fresh dairy industry, the challenges associated with demand variability are compounded by the short shelf life of the finished products and relatively long production lead times—3 to 4 days from milk pasteurization to final product. Poor production plans lead to both product waste and stock shortages, making agility and regular rescheduling critical.

The extension of the range of products in a typical dairy processing industry makes scheduling particularly difficult in order to meet mix in demand. The existence of different production lines which can carry out the same operations but which are distinguished by different production rates and the cleaning-in-place operations required between different products being processed in the same equipment makes the problem additional complex. This is further complicated by the existence of limited and shared resources (e.g., labor, utilities, etc.). There are two key shared resources in the process: operators and utilities. Different categories of operators may be available in each shift to perform the various duties involved in plant operation. If an appropriate number of operators are not available in a particular category which is needed to start a task associated with a product, the initiation of this task is delayed until the required number becomes available. Alternatively, an external production service should be used if very tight product delivery times exist (Nakhla 1995).

## 6.3 Problem Statement

In this chapter, the resource-constrained production planning problem of a multi-product semicontinuous dairy plant is addressed. More specifically, the production line under consideration produces set, stirred, or flavored yogurt. It is noted that

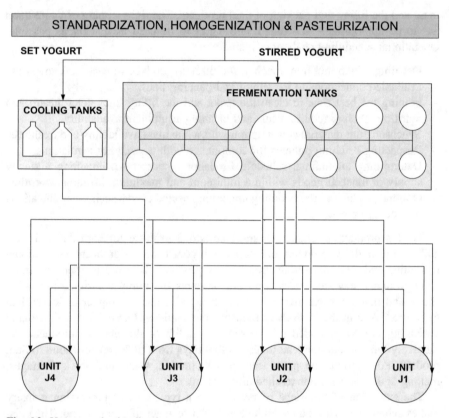

**Fig. 6.2** Yogurt production line layout

flavored yogurt is stirred yogurt with additional fruit (or other type) flavor. Thus, flavored yogurt production should pass through fruit-mixer equipment in order to perform the addition and the mixing of fruit substances. Each packing unit has a dedicated fruit-mixer. The yogurt production line consists of a set of cooling tanks (set yogurt), a set of fermentation tanks (stirred and flavored yogurt), and four packing machines. The main yogurt production line layout is illustrated in Fig. 6.2. It is clear that packing units operate in parallel. Moreover, they share common resources such as manpower.

The problem addressed here is formally defined in terms of the following items:

(i)   A known planning horizon is divided into a set of periods $n \in N$.
(ii)  A set of processing units $j \in J$ with available production time in period $n$ equal to $\omega_{jn}$.
(iii) A set of products $p \in P$ with specific production targets $\zeta_{pn}^{\text{cup}}$, inventory costs $\xi_{pn}$, production rates $\rho_{pj}$, minimum processing runs $\pi_{pjn}^{\text{min}}$, fixed $\nu_{jn}$ and variable operating costs $\theta_{pjn}$, cup weights $\eta_p^{\text{cup}}$, and external production cost

$\psi_{pn}$. $P_j$ is the subset of products that can be assigned to unit $j$ and $J_p$ is the subset of units that can produce product $p$

(iv) A set of batch recipes $r \in R$ with minimum preparation time $\tau_r$, preparation cost $\chi_{rn}$, and minimum and maximum production capacity $\mu_{rn}^{\min}$ and $\mu_{rn}^{\max}$, respectively. The subset of products that come from recipe $r$ is denoted by $P_r$.

(v) A set of product families or simply *families* $f \in F$, wherein all products are grouped into; $F_j$ is the subset of families that can be assigned to unit $j$ and $J_f$ is the subset of units that can process family $f$, while the subset of products in family $f$ is denoted by $P_f$

(vi) A set of renewable resources $k \in K$. Parameter $\varepsilon_{kfj}$ denotes the requirement of resource $k$ for processing family $f$ in unit $j$ and $E_{kn}^{\max}$ is the maximum capacity for resource $k$ in period $n$.

(vii) A (sequence-dependent) changeover operation is required in each processing unit whenever the production is changed between two different families; the required changeover time is $\gamma_{ff'j}$, while the changeover cost is $\phi_{ff'jn}$.

(viii) A (sequence-independent) setup operation is required whenever a product $p$ is assigned to a processing unit $j$; the setup time is $\delta_{pj}$.

(ix) Forbidden processing sequences for families.

The key decision variables are:

(i) The allocation of products to processing units $Y_{pjn}^p$.

(ii) The sequencing of families in every processing line $X_{ff'jn}$ and $X_{ff'jn}$;

(iii) The amount of product $p$ produced in unit $j$ ($Q_{pjn}$), and the inventory level of product $p$ at the end of planning period $n$ ($I_{pn}$);

(iv) The starting ($S_{fjn}$) and completion times ($C_{fjn}$) for every family.

So that an economic objective function typically representing total production costs is optimized.

## 6.4 Conceptual Model Design

Production planning in semicontinuous processing plants typically deals with a large number of products. Fortunately, many products appear similar characteristics. Therefore, products that share the same processing features could be treated as a product family group (family). Thus, the production planning problem under question could be *partially* focused on product families rather than on each product separately, following a similar modeling concept to Chap. 4. The definition of product families significantly reduces the size of the underlying mathematical model and, thus, the necessary computational effort without sacrificing any feasibility constraint. In the proposed approach, products belong to the same family *if and only if*: (i) they come from the same batch recipe (e.g., fermentation recipe),

(ii) they require the same labor resources, and (iii) there is no need for changeover operations among them.

When changing the production between two products that are not based on the same recipe, it is always necessary to perform changeover cleaning and/or sterilizing operations. In dairy plants, a "natural" sequence of products often exists (e.g., from the lower taste to the stronger or from the brighter color to the darker), thus the relative sequence of products within a family is usually known a priory. Therefore, when changing the production between two products of the same family, cleaning and sterilizing operations are not needed. Hence, not only the relative sequence of products, belonging to the same family, may be known but also the relative sequence of families in each processing line. In that case, different families are enumerated according to their relative position within the production day.

It should be noted that (sequence-independent) setup times, mainly depending on the cup size or product type changes, (among products of the same family) may exist, and the proposed mathematical model can treat them appropriately. Finally, it is worth mentioning that a salient feature of the proposed modeling approach is that it allows products that belong to the same family to have different: (i) processing rates, (ii) setup times/costs, (iii) minimum and maximum production runs, (iv) operating costs, and (v) inventory costs.

## 6.5   Mathematical Formulation

In the proposed mathematical framework, constraints have been grouped according to the type of decision (assignment, timing, sequencing, etc.) upon which they are imposed on. It should be emphasized that the proposed model is a crossbreed between a continuous and a discrete time representation model (Kopanos et al. 2012). More specifically, the planning horizon of interest is discretized into a number of time periods each having the duration of one production day. Then, operations within the same day are modeled using a continuous time representation (see Fig. 6.3). Mass balance is realized at the end of each production day. To facilitate the presentation of the MIP model, we use uppercase Latin letters for optimization variables and sets, and lowercase Greek letters for parameters.

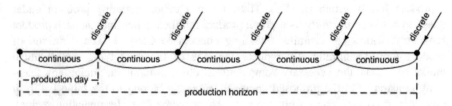

**Fig. 6.3**  Time representation

**Products Lot-Sizing and Allocation Constraints** Lower and upper bounds on the produced amounts of product $p$ are imposed by:

$$\pi_{pjn}^{\min} Y_{pjn}^p \leq Q_{pjn} \leq \pi_{pjn}^{\max} Y_{pjn}^p \quad \forall p, j \in J_p, n \tag{6.1}$$

Tighter maximum produced quantities can be estimated by:

$$\pi_{pjn}^{\max} = \left\{ \begin{array}{ll} \displaystyle\sum_{n' \geq n}^{N} \zeta_{pn'} & \text{if } \displaystyle\sum_{n' \geq n}^{N} \zeta_{pn'} < (\Lambda_{jn} - \sum_{r \in R_p} \tau_r) \rho_{pj} \\ (\Lambda_{jn} - \tau_r) \rho_{pj} & \text{if } \displaystyle\sum_{n' \geq n}^{N} \zeta_{pn'} \geq (\Lambda_{jn} - \sum_{r \in R_p} \tau_r) \rho_{pj} \end{array} \right\}$$

where $\Lambda_{jn} = \omega_{jn} - \alpha_{jn} - \beta_{jn}$, and production targets $\zeta_{pn}$ for product $p$ are given by:

$$\zeta_{pn} = \zeta_{pn}^{\text{cup}} \eta_p^{\text{cup}} \quad \forall p, n$$

Notice that $\zeta_{pn}^{\text{cup}}$ is provided by the logistics department of the company and usually reflects production targets which are based on actual products demands as well as on forecasts.

**Family Processing Time Definition** Sequencing and timing decisions need to be taken regarding families $f$ rather than products $p$, because products that belong to the same family $(p \in P_f)$ do not require changeovers among them. However, it should be noted that setup times, $\delta_{pj}$, may exist. In order to define sequencing and timing decisions for families, the definition of family processing time is introduced as follows:

$$T_{fjn} = \sum_{p \in P_f} \left( \frac{Q_{pjn}}{\rho_{pj}} + \delta_{pj} Y_{pjn}^p \right) \quad \forall f, j \in J_f, n \tag{6.2}$$

In the proposed approach, processing rates $\rho_{pj}$ are considered fixed as potential fluctuations may provoke quality problems (Soman et al. 2004).

**Families Allocation Constraints** A family $f$ is assigned to a processing unit $j$ in a production day $n$ if at least one product $p \in P_f$ that belongs to this family, is processed on this unit during the same production day:

$$Y_{fjn} \geq Y_{pjn}^p \quad \forall f, p \in P_f, j \in J_f, n \tag{6.3}$$

**Families Sequencing and Timing Constraints** Constraint sets (6.4) and (6.5) state that if a family $f$ is allocated to processing unit $j$ in period $n$, (i.e., $Y_{fjn} = 1$) at most one family $f'$ is processed before and/or after it, respectively.

$$\sum_{f' \neq f, f' \in F_f} X_{f'fjn} \leq Y_{fjn} \quad \forall f, j \in J_f, n \tag{6.4}$$

$$\sum_{f' \neq f, f' \in F_f} X_{ff'jn} \leq Y_{fjn} \quad \forall f, j \in J_f, n \tag{6.5}$$

The total number of active sequencing binary variables $X_{ff'jn}$ plus the unit utilization binary variable $V_{jn}$ should be equal to the total number of active allocation binary variables $Y_{fjn}$ in a processing unit $j$ at period $n$, according to constraint set (6.6). For instance, if three families are assigned to a unit $j$ then two sequencing variables will be active.

$$\sum_{f \in F_j} \sum_{f' \neq f, f' \in F_j} X_{ff'jn} + V_{jn} = \sum_{j \in F_j} Y_{fjn} \quad \forall j, n \tag{6.6}$$

Constraint set (6.7) ensures that the processing unit $j$ is used in period $n$, (i.e., $V_{jn} = 1$) if at least one family $f$ is assigned to it in this period (i.e., $Y_{fjn} = 1$). Note that no lower bound on the binary variable $V_{jn}$ is necessary as far as a cost term (related to the unit utilization), is included into the objective function, thus enforcing $V_{jn}$ to zero.

$$V_{jn} \geq Y_{fjn} \quad \forall f, j \in J_f, n \tag{6.7}$$

Constraint set (6.8) states that the starting time of a family $f'$, $S_{f'jn}$ that follows another family $f$ in a processing line $j$ in period $n$, (i.e., $X_{ff'jn} = 1$) is greater than the completion time of family $f$, $C_{fjn}$, plus the necessary changeover time $\gamma_{ff'j}$ between these families. Note that the big-M parameter $M_{jn}$ can be set equal to $(\omega_{jn} - \beta_{jn})$, where $\omega_{jn}$ is the available production time horizon and $\beta_{jn}$ corresponds to the daily plant shutdown time.

$$C_{fjn} + \gamma_{ff'j} \leq S_{f'jn} + M_{jn}(1 - X_{ff'jn}) \quad \forall f, f' \neq f, j \in (J_f \cap J_{f'}), n \tag{6.8}$$

Obviously,

$$S_{fjn} = C_{fjn} - T_{fjn} \quad \forall f, j \in J_f, n \tag{6.9}$$

### Families Completion Times Lower and Upper Bounds

Constraints (6.10) and (6.11) impose a lower and upper bound on each family completion time, $C_{fjn}$, respectively. More specifically, the completion time has to be greater than the daily plant setup time, $\alpha_{jn}$, plus the minimum time $\tau_r$ for preparing the batch recipe (e.g., fermentation recipe) $r$, plus the processing time, $T_{fjn}$, and the changeover time, $\gamma_{ffj}$, for changing the production to family $f$. An additional unit preparation time $o_{jn}$ is also taken into account. This time stands for the additional preparation time of a processing unit $j$ due to potential maintenance or other technical reasons. Additionally, the release batch recipe time $\sigma_{rn}$ is also considered. In order to commence the production of a batch recipe $r$, all recipes ingredients need to be present. Otherwise, the production of the batch recipe $r$ will be postponed until the arrival of its missing substances.

$$C_{fjn} \geq (\alpha_{jn} + \max[o_{jn}, \sigma_{rn}] + \tau_r)Y_{fjn} + T_{fjn}$$
$$+ \sum_{f' \neq f, f' \in F_j} \gamma_{f'fj} X_{f'fjn} \quad \forall f, r \in R_f, j \in J_f, n \tag{6.10}$$

Constraint set (6.11) ensures that the completion time of each family is smaller than the daily production time horizon $\omega_{jn}$ minus the daily plant shutdown time $\beta_{jn}$. Production line shutdown is realized on a daily basis, as a typical production policy to guarantee high quality of final products and to comply with hygienic standards.

$$C_{fjn} \leq (\omega_{jn} - \beta_{jn})Y_{fjn} \quad \forall f, j \in J_f, n \tag{6.11}$$

**Batch Recipe Stage Constraints** Batch recipe stage (e.g., fermentation and pasteurization) constraints must be included into the mathematical model in order to guarantee a feasible production plan in yogurt production lines. The cumulative produced quantity of products $p \in P_r$ should be greater than the minimum produced batch recipe amount (e.g., in the pasteurization and fermentation stages) $\mu_{rn}^{\min}$ and lower than the maximum production capacity $\mu_{rn}^{\max}$:

$$\mu_{rn}^{\min} Y_{rn}^R \leq \sum_{p \in P_r} \sum_{j \in J_p} Q_{pjn} \leq \mu_{rn}^{\max} Y_{rn}^R \quad \forall r, n \tag{6.12}$$

Constraint set (6.13) ensures that a batch recipe $r$ is produced in period $n$, (i.e., $Y_{rn}^R = 1$), if at least one $f \in F_r$ is processed on a processing unit $j$ in the same period $n$ (i.e., $Y_{fjn} = 1$).

$$Y_{rn}^R \geq \sum_{j \in J_f} Y_{fjn} \quad \forall r, f \in F_r, n \tag{6.13}$$

**Tightening Constraints** In order to reduce the computational effort, constraints (6.14) can further tighten the linear relaxation of the proposed mathematical model by imposing an upper bound on the total processing time for every processing line $j$ in each period $n$.

$$\sum_{f \in F_j} T_{fjn} \leq (\omega_{jn} - \alpha_{jn} - \beta_{jn} - \min_{r \in R_f}[\tau_r])V_{jn}$$
$$- \sum_{f \in F_j} \sum_{f' \neq f, f' \in F_j} \gamma_{ff'j} X_{ff'jn} \quad \forall j, n \tag{6.14}$$

Note that by incorporating constraint set (6.14) into the mathematical formulation, constraint set (6.7) can be omitted, thus further reducing the model size.

**Products Mass Balance Constraints** The total quantity of product $p$ produced on the plant (internal production) in period $n$, $Q_{pn}^{int}$, is given by

$$Q_{pn}^{int} = \sum_{j \in J_p} Q_{pjn} \quad \forall p, n \tag{6.15}$$

At this point, it is worth pointing out that demand satisfaction is of great importance in the dairy industry. Inability to satisfy customer demand on time may result in losses of the market share, competitive advantage, increase customers disappointment, etc. Therefore, the full demand satisfaction is desired. Constraints (6.16) enforce full demand satisfaction. Figure 6.4 presents a network representation of the production planning problem under question. The inventory $I_{pn}$ of product $p$ is the summation of the previous period inventory, $I_{pn-1}$, plus the total internal, $Q_{pn}^{int}$, and external production, $Q_{pn}^{ext}$, minus the production target, $\zeta_{pn}$, in the current period $n$:

$$I_{pn} = I_{pn-1} + Q_{pn}^{int} + Q_{pn}^{ext} - \zeta_{pn} \quad \forall p, n \tag{6.16}$$

External production usually expresses production targets that exceed the production capacity of the dairy plant. It can be realized in an affiliated production facility, if one exists, otherwise, it represents the unsatisfied demand. A high penalty cost for external production will enforce the MIP model to generate solutions that comply with full demand satisfaction by internal production. Obviously, the external production of product $p$ in production day $n$ cannot be greater than the product demand at the same production day:

$$Q_{pn}^{ext} \leq \zeta_{pn} \quad \forall p, n \tag{6.17}$$

Constraint set (6.18) is added to the proposed MIP model if product safety stocks are desired. If product-dependent storage limitations exist, constraint set (6.19) is used. Otherwise, constraint set (6.20) is included to account for the total plant storage capacity.

$$I_{pn} \geq \text{product safety stock} \quad \forall p, n \tag{6.18}$$

$$I_{pn} \leq \text{product storage capacity} \quad \forall p, n \tag{6.19}$$

**Fig. 6.4** Network representation of the production planning problem

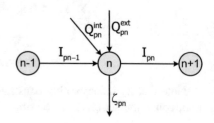

$$\sum_{p} I_{pn} \leq \text{total plant storage capacity} \quad \forall n \tag{6.20}$$

**Objective Function** The objective function to be minimized is the total cost including several factors such as: (i) inventory costs, (ii) operating costs, (iii) batch recipes preparation costs, (iv) unit utilization costs, (v) families changeover costs, and (vi) external production costs, as follows:

$$
\begin{aligned}
\min \quad & \sum_{p}\sum_{n} \xi_{pn} I_{pn} + \sum_{p}\sum_{j\in J_{p}}\sum_{n} \frac{\theta_{pjn}}{\rho_{pj}} Q_{pjn} + \sum_{r}\sum_{n} \chi_{rn} Y_{rn}^{R} \\
& + \sum_{j}\sum_{n} v_{jn} V_{jn} + \sum_{f}\sum_{f'\neq f}\sum_{j\in(J_{f}\cap J_{f'})}\sum_{n} \phi_{ff'jn} X_{ff'jn} \\
& + \sum_{p}\sum_{n} \psi_{pn} Q_{pn}^{\text{ext}}
\end{aligned}
\tag{6.21}
$$

In a dairy plant, final yogurt products are kept at low temperatures, thus resulting to a significant inventory cost (mainly due to high energy requirements), which should be considered in the optimization procedure. Moreover, inventory costs should include shelf life issues. Roughly speaking, the lower the shelf life of a product, the higher its inventory cost. Operating costs mainly include labor and energy costs plus costs due to material losses. The fermentation recipe cost account for all costs associated with the preparation of each fermentation recipe. The unit utilization cost basically stands for the shutdown cleaning operation cost plus the initial unit setup cost. Changeover costs correspond to cleaning and/or sterilization operations. Finally, external production costs reflect the penalty cost of producing the requested production targets to an affiliated production facility. The nature of this cost is more qualitative than quantitative. A high external production cost will enforce demand satisfaction by internal production. In this case, external production will appear only if the production targets are higher than the production capacity of the plant. In other words, a full demand satisfaction by internal production is indirectly favored. It is worthy mentioning that since full demand satisfaction is imposed, the minimization of total costs is identical to the maximization of total profit.

## 6.5.1   Extension to Renewable Resources Constraints

In most industrial environments, resource limitations often constitute a crucial part of the production planning problem. By neglecting potential resource constraints in the optimization procedure, there is no guaranteeing that a feasible production plan will not be obtained.

Roughly speaking, resources could be mainly classified into *nonrenewable* and *renewable*. Nonrenewable resources do not recover their capacity after the completion of the tasks that consumed them. For instance, raw materials and intermediate products can be considered as nonrenewable resources. On the other hand, renewable resources recover their capacity after the completion of the tasks that used them. Renewable resources like manpower are called discrete renewable resources, while resources such as utilities (e.g., electricity, vapor, cooling water, etc.) are usually referred to as continuous renewable resources.

In the dairy processing industry, the available manpower usually constitutes the major resource limitation. This is the case in the plant under consideration where a limited number of employees are available during each production day. The modeling approach of labor resources in the present work follows similar modeling concepts to the recent contribution of Marchetti and Cerdá (2009).

**Basic Conditions for Modeling Resources Constraints** By definition, a family $f'$ that is overlapping the starting time of family $f$ must satisfy the following conditions:

(A) It should demand some resource $k$ also required by family $f$.
(B) It is assigned to a processing unit different from the one that is allocated to family $f$.
(C) It starts before or exactly at the time that family $f$ starts being processed (i.e., $S_{f'j'n} \leq S_{fjn}$).
(D) It should end after the starting time of family $f$ (i.e., $C_{f'j'n} > S_{fjn}$).

An illustrative example of the basic overlapping conditions is shown in Fig. 6.5. Note that family $f$: (i) is overlapped by family $f'$ and (ii) is overlapping family $f''$.

**Sequencing Constraints for Families Assigned to Different Units** Constraints (6.22) to (6.24) are included into the MIP model to ensure the families' relative sequencing related to their starting times; condition (C). Global sequencing binary

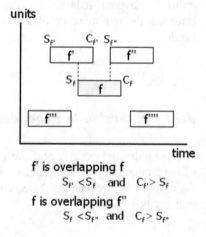

**Fig. 6.5** Illustrative example for overlapping conditions

variables $\bar{X}_{f'j'fjn}$ are introduced for each pair of families $f'$ and $f$ that are assigned to different units. When family $f'$, which is allocated to processing unit $j'$, starts before family $f$, which is allocated to processing unit $j \neq j'$, the binary variable, $\bar{X}_{f'j'fjn}$, is active (i.e., $\bar{X}_{f'j'fjn} = 1$). A very small number $\lambda$ is added in constraint set (6.24) to effectively cope with the case when two families' $f$ and $f'$ start at the same time point. In this case, it is assumed that family $f' < f$ starts slightly before family $f$. In other words, the family with the lower index begins first. Note that, if both or one of the families $f$ and $f'$ is not assigned to unit $j$ or unit $j' \neq j$, respectively, (i.e., $Y_{fjn} = 0$ and/or $Y_{f'j'n} = 0$) constraints (6.22)–(6.24) become redundant.

$$S_{f'j'n} - S_{fjn} \leq M_{jn}(1 - \bar{X}_{f'j'fjn}) + M_{jn}(2 - Y_{f'j'n} - Y_{fjn})$$
$$\forall f, f', j \in J_f, j' \in J_{f'}, n : j' \neq j \tag{6.22}$$

$$S_{fjn} - S_{f'j'n} \leq M_{jn}\bar{X}_{f'j'fjn} + M_{jn}(2 - Y_{f'j'n} - Y_{fjn})$$
$$\forall f, f' \geq f, j \in J_f, j' \in J_{f'}, n : j' \neq j \tag{6.23}$$

$$S_{fjn} - S_{f'j'n} + \lambda \leq M_{jn}\bar{X}_{f'j'fjn} + M_{jn}(2 - Y_{f'j'n} - Y_{fjn})$$
$$\forall f, f' < f, j \in J_f, j' \in J_{f'}, n : j' \neq j \tag{6.24}$$

**Families Overlapping Constraints** In order to derive the mathematical expression for the overlapping condition (D), an auxiliary overlapping binary variable, $Z_{f'j'fjn}$ is defined. This variable is active (i.e., $Z_{f'j'fjn} = 1$) whenever family $f'$ is completed after the starting time of family $f$, as constraint set (6.25) states. If both or one of the families $f$ and $f'$ is not assigned to unit $j$ or unit $j' \neq j$, respectively, (i.e., $Y_{fjn} = 0$ and/or $Y_{f'j'n} = 0$) constraint set (6.25) becomes redundant.

$$C_{f'j'n} - S_{fjn} \leq M_{jn}Z_{f'j'fjn} + M_{jn}(2 - Y_{f'j'n} - Y_{fjn})$$
$$\forall f, f', j \in J_f, j' \in J_{f'}, n : j' \neq j \tag{6.25}$$

Whenever the RHS of constraint set (6.25) is positive (i.e., $C_{f'j'n} - S_{fjn} > 0$), condition (D) is satisfied. Note that, in this case, the auxiliary overlapping binary variable $Z_{f'j'fjn}$ is enforced to take the value of 1.

It can be easily proven that if two families $f$ and $f'$ are running in parallel, the Boolean condition $(C_{f'j'n} > S_{fjn}) \wedge (C_{fjn} > S_{f'j'n})$ is satisfied and, therefore, $Z_{fjf'j'n} + Z_{f'j'fjn} = 2$. However, it is important to keep in mind that not every family $f'$ satisfying the necessary condition $Z_{fjf'j'n} + Z_{f'j'fjn} = 2$ is an overlapping family, but only those families running at the starting time $S_{fjn}$. Given the condition $Z_{fjf'j'n} + Z_{f'j'fjn} = 2$, a global sequencing variable $\bar{X}_{f'j'fjn}$ is required to decide which family ($f$ or $f'$) overlaps the other one. If $Z_{fjf'j'n} + Z_{f'j'Vfjn} = 2$ and $\bar{X}_{f'j'fjn} = 1$, then family $f'$ is overlapping family $f$; according to constraints (6.26) and the overlapping binary variable $W_{f'j'fjn}$ takes the value of 1 in this case.

$$W_{f'j'fjn} \geq Z_{f'j'fjn} + \bar{X}_{f'j'fjn} - Y_{f'j'n}$$
$$\forall f, f', j \in J_f, j' \in J_{f'}, n : j' \neq j \tag{6.26}$$

**Families Resources Capacity** Constraint set (6.27) does not allow family $f$ to start being processed if the maximum resource capacity $E_{kn}^{\max}$ is reached. Thus, resource overloads, which result in infeasible solutions, are avoided.

$$\varepsilon_{kfj} Y_{fjn} + \sum_{f' \in F_k} \sum_{j' \neq j, j' \in J_{f'}} \varepsilon_{kf'j'} W_{f'j'fjn} \leq E_{kn}^{\max}$$
$$\forall k, f \in F_k, j \in J_f, n \tag{6.27}$$

The proposed modeling approach is able to tackle problems of multiple renewable resources constraints; either discrete (such as manpower) or continuous (such as utilities) types. Note that unit-dependent resource requirements can be also considered explicitly by the proposed set of resource constraints.

**Modified Objective Function** To express the effect of labor resources, a non-quantitative managerial term is added in the overall objective function. This term expresses the number of employees that are working simultaneously. It is preferable to keep this number as low as possible in order to use the remaining manpower in other tasks or to preserve them in case of the occurrence of an unexpected event. More importantly, the less the manpower used, the lower the possibilities for manpower errors, and bad coordination and the higher the production flexibility. Finally, an auxiliary penalty term is also introduced to explicitly take account of potential resource constraints.

$$
\begin{aligned}
\min \sum_p \sum_n \xi_{pn} St_{pn} &+ \sum_p \sum_{j \in J_p} \sum_n \frac{\theta_{pjn}}{\rho_{pj}} Q_{pjn} + \sum_r \sum_n \chi_{rn} Y_{rn}^R \\
&+ \sum_j \sum_n \nu_{jn} V_{jn} + \sum_f \sum_{f' \neq f} \sum_{j \in (J_f \cap J_{f'})} \sum_n \phi_{ff'jn} X_{ff'jn} \\
&+ \sum_p \sum_n \psi_{pn} Q_{pn}^{ext} + \sum_f \sum_{f'} \sum_{j \in J_f} \sum_{j' \neq j, j' \in J_{f'}} \sum_n W_{f'j'fjn} \\
&+ \sum_f \sum_{f'} \sum_{j \in J_f} \sum_{j' \neq j, f' \in J_{f'}} \sum_n Z_{f'j'fjn}
\end{aligned}
\tag{6.28}
$$

## 6.6   Industrial Case Studies

In this section, a number of complex real-world production planning problems in the yogurt production line of the KRI-KRI diary production facility, located in Northern Greece, are considered (Kopanos et al. 2012). The facility under study consists of four packing units (J1–J4), working in parallel and producing a total of

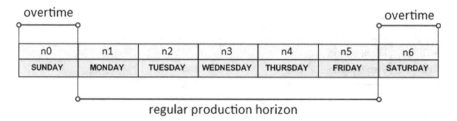

**Fig. 6.6** Production planning horizon

93 yogurt products that are grouped into 23 families (F01–F23). Real data have been slightly modified due to confidentiality issues.

The production time horizon in the underlying yogurt production facilities is usually 1 week (Nakhla 1995). The regular production is performed from Monday to Friday and overtime may be permitted on Sunday and/or on Saturday (see Fig. 6.6). The duration of each production period is 24 h. The daily scheduled cleaning operations of the plant $\beta_{jn}$ last 2 h. The total plant setup time $\alpha_{jn}$ is 3 h representing the necessary time for the completion of production stages before the fermentation stage (pasteurization, homogenization, etc.). Product demand data represent packing stage production targets have been provided from the logistics department of the plant. Demands quantities and due dates are based on product orders from Sunday to Tuesday of the following week as well as on forecasts. Demand due dates are given for packed final products (subtracting the necessary final cold storage and quality control time, which varies between 2 and 5 days).

The main processing data for final products and families (i.e., classification of products to families, cup weights, inventory costs, minimum production runs, packing rates, and families changeover times and costs) can be found in Appendix C. Table 6.1 provides: (i) the main data for each fermentation recipe including the minimum fermentation time (stirred yogurt) or the minimum cooling time (set yogurt) for preparing each fermentation recipe $r$, (ii) the recipe preparation cost, and (iii) the set of product families $f \in F_r$ that share the same fermentation recipe. Table 6.2 illustrates the relative sequence of families in a production day. The minimum produced quantity of any fermentation recipe $\mu_r^{min}$, due to pasteurization and fermentation stage operability issues, is 1200 kg. The minimum packing time for any family is equal to 0.5 h. Manpower requirements for each family are shown in Table 6.3.

The variable operating cost, $\theta_{pjn}$, expresses mainly labor and utilities costs of the packing stage. This is equal to 1000 €/h for any packing unit, during a regular production day (weekdays) and 10,000 €/h in overtime periods (Saturday and Sunday); this actually reflects the industrial policy to keep the production facility closed during the weekend. Moreover, the cost for the production of a fermentation recipe in overtime periods (weekend) is taken twice the cost for producing it into a regular production period (weekdays). The fixed utilization and cleaning packing unit cost, $\nu^{unit}$, for each packing line is 1000 € in a regular production day and

**Table 6.1**  Main data for recipes

| Recipe | Process type | Preparation time (h) | Cost (€) | Families |
|---|---|---|---|---|
| R01 | Fermentation | 4.75 | 545 | F04, F11 |
| R02 | Fermentation | 4.50 | 540 | F05, F12 |
| R03 | Fermentation | 8.25 | 565 | F13 |
| R04 | Fermentation | 7.75 | 555 | F14, F15 |
| R05 | Fermentation | 5.25 | 525 | F20 |
| R06 | Fermentation | 7.25 | 565 | F19, F21, F22 |
| R07 | Fermentation | 8.75 | 625 | F06, F07, F18 |
| R08 | Cooling | 1.50 | 505 | F01, |
| R09 | Cooling | 1.50 | 510 | F02 |
| R10 | Cooling | 1.50 | 515 | F03 |
| R11 | Fermentation | 8.75 | 625 | F08, F09, F10, F16, F17 |
| R12 | Fermentation | 8.75 | 600 | F23 |

**Table 6.2**  Families relative sequences in a production day per packing line

| Unit | Families relative sequence |
|---|---|
| J1 | F20=>F21=>F22 |
| J2 | F12=>F11=>F19=>F18=>F13=>F14=>F15=>F16=>F17 |
| J3 | F01=>F02=>F03=>F05=>F04=>F08=>F09=>F10=>F06=>F07 |
| J4 | F08=>F09=>F10=>F06=>F23 |

**Table 6.3**  Manpower requirements for families in every packing unit

| Family | Units | | | | Family | Units | | | |
|---|---|---|---|---|---|---|---|---|---|
| | J1 | J2 | J3 | J4 | | J1 | J2 | J3 | J4 |
| F01 | | | 5 | | F13 | | 2 | | |
| F02 | | | 5 | | F14 | | 2 | | |
| F03 | | | 5 | | F15 | | 2 | | |
| F04 | | | 4 | | F16 | | 2 | | |
| F05 | | | 4 | | F17 | | 2 | | |
| F06 | | | 4 | 4 | F18 | | 2 | | |
| F07 | | | 4 | | F19 | | 2 | | |
| F08 | | | 4 | 4 | F20 | 2 | | | |
| F09 | | | 4 | 4 | F21 | 3 | | | |
| F10 | | | 4 | 4 | F22 | 2 | | | |
| F11 | | 2 | | | F23 | | | | 3 |
| F12 | | 2 | | | | | | | |

5000 € in overtime periods. In order to avoid the undesirable case of external production, a high external production penalty cost $\psi_{pn}$ equal to 30 €/kg is imposed. The plant employs maximum 12 workers.

All case studies have been solved to optimality in an Intel Core 2 Quad 2.84 GHz with 3.5 GB RAM using CPLEX 11 under standard configurations via a GAMS 22.8 interface. The detailed production plan for every case study is reported in Appendix D.

**Case Study I** Production targets for this case study are provided in Appendix C. There is no minimum safety stock for any period. The mathematical model consists of 17,709 equations, 10,734 binary variables, and 2664 continuous variables. The optimal solution was reached in just 142 CPUs corresponding to a total cost equal to 315,627 €.

Figure 6.7 presents the production plan for families as well as the manpower profile over the entire planning horizon of interest. The solution does not indicate production over the weekends, thus minimizing total costs. The total cost breakdown is shown in Fig. 6.8. Note that the total inventory cost represents approximately 41% of the total costs. The total changeovers cost reflects the 21% of the total costs. The inventory cost profile for each production day is illustrated in Fig. 6.9. Thursday is the day with the higher inventory cost representing about 42.3% of the total inventory cost. On the other hand, Monday is the day with the lower inventory cost contribution representing 12.4% of the total inventory cost. It is worth noting that the proposed solution does not lead to any external production.

**Case Study II** This case study considers the unexpected case of the absence of an employee (illness, etc.) from Wednesday to Friday. That means that the maximum manpower capacity is reduced from 12 to 11 workers. Therefore, a new production

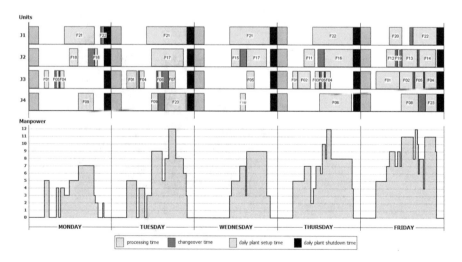

**Fig. 6.7** Case study I: production plan and manpower profile

**Fig. 6.8** Case study I: breakdown of total cost (€)

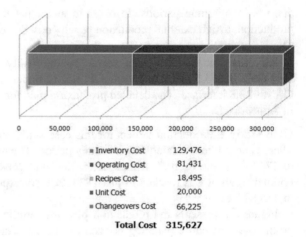

| | |
|---|---|
| ■ Inventory Cost | 129,476 |
| ■ Operating Cost | 81,431 |
| ■ Recipes Cost | 18,495 |
| ■ Unit Cost | 20,000 |
| ■ Changeovers Cost | 66,225 |
| **Total Cost** | **315,627** |

**Fig. 6.9** Case study I: total inventory cost per production period (€)

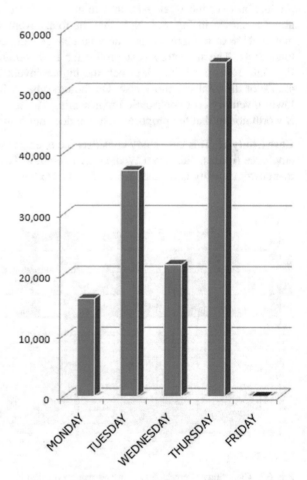

plan should be generated from Wednesday to Friday since the previous one (Case Study I), as indicated in Fig. 6.7, becomes infeasible (see also the manpower profile on Thursday and on Friday).

The production replanning problem from Wednesday to Friday using the updated manpower capacity and actual production targets is therefore considered using the proposed model. The mathematical model consists of 8901 equations, 5367 binary variables, and 1425 continuous variables. The optimal solution was reached in 516 CPUs leading to a total cost of 222,847 €.

The proposed production plan for all families and the manpower profile over the whole planning horizon is depicted in Fig. 6.10. Again, there is no need for external production. Figure 6.11 illustrates the total cost breakdown. The total inventory cost and the total changeovers cost represent approximately 40% and 24% of the total cost, respectively. The inventory cost profile for each production day is illustrated in Fig. 6.12. As expected, the solution leads to an increase in the inventory cost on Wednesday and Thursday comparing to the initial production plan (Case Study I). This is due to the fact that the number of available employees is decreased thus resulting in a production capacity decrease. Note that the inventory cost generated on Thursday is approximately three times higher than the corresponding cost on Wednesday.

**Case Study III** A salient feature of the dairy industry is that customers usually confirm (i.e., change) their order quantities a few days prior to dispatch. This case study considers the case where production targets levels change due to potential orders cancelation, arrival of new orders and modification of old orders quantities. New production targets from Tuesday to Friday arrived from the logistics department on Wednesday night (see Table 6.4).

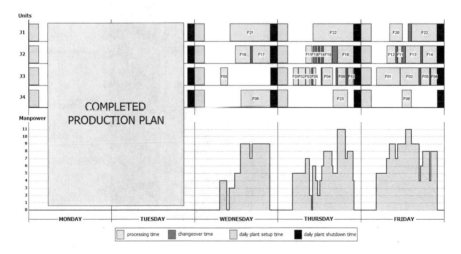

**Fig. 6.10** Case study II: production plan and manpower profile

**Fig. 6.11** Case study II:
breakdown of total cost (€)

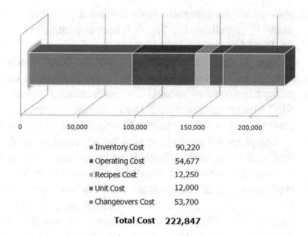

| | |
|---|---|
| ▪ Inventory Cost | 90,220 |
| ▪ Operating Cost | 54,677 |
| ▪ Recipes Cost | 12,250 |
| ▪ Unit Cost | 12,000 |
| ▪ Changeovers Cost | 53,700 |
| **Total Cost** | **222,847** |

**Fig. 6.12** Case study II: total
inventory cost per production
period (€)

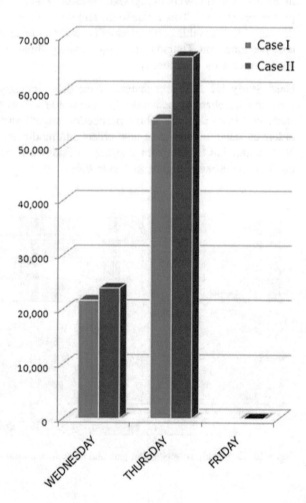

**Table 6.4** Case study III. Updated production targets (cups)

| Product | Old production targets | | New production targets | | Order modification type |
|---------|------------------|--------|------------------|--------|------------------------|
|         | Thursday | Friday | Thursday | Friday | |
| P01 |        |        | 1402  |        | New arrival |
| P04 |        | 14,001 |       | 11,474 | Quantity modified |
| P05 |        | 5480   |       | 7985   | Quantity modified |
| P08 |        | 4000   |       |        | Cancelled |
| P11 |        | 715    |       | 946    | Quantity modified |
| P13 |        |        | 1628  |        | New arrival |
| P16 |        | 3715   |       | 4155   | Quantity modified |
| P17 |        |        | 1928  |        | New arrival |
| P21 |        | 1620   |       | 2220   | Quantity modified |
| P22 |        | 1380   |       | 1790   | Quantity modified |
| P24 | 4193   |        |       |        | Cancelled |
| P25 | 14,974 |        | 13,792 |       | Quantity modified |
| P28 |        |        |       | 2146   | New arrival |
| P33 |        | 4057   |       | 3002   | Quantity modified |
| P41 |        | 1172   |       | 2276   | Quantity modified |
| P44 |        | 2019   |       | 1626   | Quantity modified |
| P58 | 3188   |        | 1985  |        | Quantity modified |
| P64 |        |        |       | 1856   | New arrival |
| P72 | 2040   |        |       |        | Cancelled |
| P80 |        |        |       | 120    | New arrival |
| P81 |        |        |       | 240    | New arrival |

Therefore, a new production plan, considering the new production targets, should be generated from Tuesday to Friday. The initial production plan is that of Case Study I. The optimal production plan was obtained in 242 CPUs leading to a total cost of 179,793 €. An external production of 3290 kg for product P87 is observed, since the production capacity of the dairy plant is not able to achieve full demand satisfaction. Figure 6.13 illustrates the production plan for families and the manpower profile over the entire planning horizon. An inventory cost equal to 68,161 € is generated on Thursday while this figure for Friday is 1530 €. The canceled product order for P72 reflects the inventory cost of Friday, since P72 had been already produced in advance on Monday (see Table D.1 in Appendix D).

Figure 6.14 illustrates the total cost breakdown. The external production cost is not included in the objective function since it reflects a penalty cost and not a real cost term.

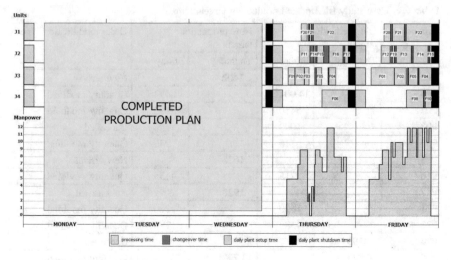

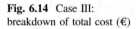

**Fig. 6.13**  Case III: production plan and manpower profile

**Fig. 6.14**  Case III:
breakdown of total cost (€)

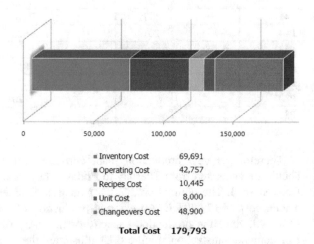

| | |
|---|---|
| ▪ Inventory Cost | 69,691 |
| ▪ Operating Cost | 42,757 |
| ▪ Recipes Cost | 10,445 |
| ▪ Unit Cost | 8,000 |
| ▪ Changeovers Cost | 48,900 |
| **Total Cost** | **179,793** |

## 6.7  Concluding Remarks

In this chapter, an MIP framework for the resource-constrained production planning problem in semicontinuous processing industries (e.g., dairy industries) has been presented. Quantitative, as well as qualitative optimization goals, are included in the proposed model. Renewable resource limitations are appropriately taken into account. Moreover, the MIP formulation has been extended to deal with unexpected events such as the absence of an employee, product order cancelation or modification, etc. The properly treatment of uncertainty in semicontinuous industries is of great importance since unpredicted events take place very frequently. Food processing industries involve the production of perishable products, therefore, strategies of building up inventories are inappropriate because they compromise the freshness, the quality, and the selling price of the final products. Therefore, and as illustrated in Case studies II and III, production replanning should be done *online* after the occurrence of an unexpected event.

The presented MIP model aims at being the core element of a computer-aided advanced planning system in order to facilitate decision-making in related industrial environments. More specifically, the proposed approach can help users analyze plans and schedules, run what-if analysis, compare scenarios, balance the optimization of multiple goals, modify the recommended solution, and determine whether a modification violates any constraints. The results indicate the best possible production plans and schedules to maximize profitability and customer service, while taking into account the full set of operating costs and constraints, from inventory carrying and changeover costs to equipment management and labor resource availability. The proposed planning model delivers value beyond plan feasibility and schedule optimization. It may also serve as a tool for negotiations between the manufacturing and supply chain departments, allowing them to collaborate more easily to find the best balance between inventory levels and operational efficiency. Furthermore, it can provide the basis to analyze the impact of new production plans on manufacturing efficiency, and scheduling decisions on inventory levels and demand satisfaction.

In lack of computer-aided production planning tools, empirical production plans are usually sent to the plant floor in dairy processing industries; thus thwarting the lucrative performance of the production facility. It should be pointed out that it may be difficult to directly quantify the benefits of the proposed MIP framework because the pre-computer situation is not usually known in detail, so there is no sufficient basis for comparison. However, this single fact is an excellent argument in favor of computer-aided production planning as discussed by Jakeman (1994): *If you do not know how well you are doing, how can you improve your performance?*

## 6.8  Nomenclature

**Indices/Sets**

| | |
|---|---|
| $f, f' \in F$ | Product families (families) |
| $f, f' \in J$ | Product batches (batches) |
| $k \in K$ | Renewable resources |
| $n \in N$ | Planning time periods |
| $p \in P$ | Products |
| $r \in R$ | Batch recipes (recipes) |

**Subsets**

| | |
|---|---|
| $F_j$ | Families $f$ that can be processed in unit $j$ |
| $F_k$ | Families $f$ that share the same renewable resource $k$ |

$F_r$    Families $f$ that have the same recipe origin $r$
$J_f$    Available units $j$ to process family $f$
$J_p$    Units $j$ that can process product $p$
$P_f$    Products $p$ that belong to the same family $f$
$P_r$    Products $p$ that have the same recipe origin $r$
$R_f$    Recipe origin $r$ for family $f$
$R_j$    Recipes $r$ that can be processed in unit $j$
$R_p$    Product $p$ that comes from recipe $r$

**Parameters**

$\alpha_{jn}$    Daily opening setup time for every unit $j$ in period $n$ (e.g., accounts for the pasteurization and homogenization stages)
$\beta_{jn}$    Daily shutdown time for every unit $j$ in period $n$ (e.g., cleaning of yogurt production line for hygienic and quality reasons)
$Y_{ff'j}$    Changeover time between family $f$ and $f0$ in unit $j$ (e.g., accounts for cleaning and sterilizing operations)
$\delta_{pj}$    Setup time for product $p$ on unit $j$
$\varepsilon_{kfj}$    Renewable resource $k$ requirements for family $f$ when processed in unit $j$; in the current study corresponds to the number of workers
$E_{kn}^{max}$    Maximum total capacity of renewable resource $k$ at period $n$
$\zeta_{pn}$    Maximum total capacity of renewable resource $k$ at period $n$
$\zeta_{pn}^{cup}$    Production target for product $p$ in period $n$ (in cups)
$\eta_{pn}^{cup}$    Cup weight for product $p$
$\theta_{pjn}$    Variable operating cost for processing product $p$ in unit $j$ in period $n$ (e.g., includes labor and utilities costs)
$\lambda$    A very small number (0.001)
$\Lambda_{jn}$    $= \omega_{jn} - \alpha_{jn} - \beta_{jn}$
$M_{jn}$    $= \omega_{jn} - \beta_{jn}$
$\mu_{rn}^{max}$    Maximum production capacity of recipe $r$ in period $n$
$\mu_{rn}^{min}$    Minimum produced quantity of recipe $r$ in period $n$ (e.g., accounts for pasteurization and fermentation tanks capacity restrictions)
$v_{jn}$    Fixed cost for utilizing processing unit $j$ in period $n$
$\xi_{pn}$    Inventory cost for product $p$ at time $n$
$o_{jn}$    Additional unit preparation time for processing unit $j$ in period $n$
$\pi_{pjn}^{max}$    Maximum production run for product $p$ in unit $j$ in period $n$
$\pi_{pjn}^{min}$    Minimum production run for product $p$ in unit $j$ in period $n$
$\rho_{pj}$    Processing rate for product $\rho$ in unit $j \in J_p$
$\sigma_{rn}$    Release time for recipe $r$ in period $n$
$\tau_r$    Minimum time for preparing recipe $r$; (e.g., for producing stirred yogurt products stands for the minimum fermentation time, while for set yogurt
$\phi_{ff'jn}$    Changeover cost between family $f$ and $f0$ in unit $j$ in period $n$ (e.g., accounts for cleaning and sterilizing operations)
$\chi_{rn}$    Cost for producing recipe $r$ in period $n$

$\psi_{pn}$     External production penalty cost for product $p$ in period $n$

$\omega_{jn}$     Physical available processing time in unit $j$ at period $n$

## Continuous Variables

$C_{fjn}$    Completion time for family $f$ in unit $j$ in period $n$

$I_{pn}$     Inventory of product $p$ at time $n$

$Q_{pjn}$    Produced amount of product $p$ in unit $j$ in period $n$

$Q_{pn}^{ext}$    External production of product $p$ in period $n$

$Q_{pn}^{Int}$    Total internal production of product $p$ in period $n$

$S_{fjn}$     Starting time for family $f$ in unit $j$ in period $n$

$T_{fjn}$     Processing time for family $f$ in unit $j$ in period $n$

## Binary Variables

$V_{jn}$     = 1, If unit $j$ is used in period $n$

$W_{f'j'fjn}$    = 1, If family $f'$, assigned to unit $j'$ in period $n$, is overlapped by family $f$, assigned to unit $j \neq j'$ in the same period $n$

$X_{ff'jn}$    = 1, If family $f$ is processed exactly before family $f'$, when both are assigned to the same unit $j$ in the same period $n$

$\bar{X}_{f'j'fjn}$    = 1, If family $f'$, assigned to unit $j'$ in period $n$, starts processing before family $f$, assigned to unit $j \neq j'$ in the same period $n$

$Y_{fjn}$     = 1, If family $f$ is assigned to unit $j$ in period $n$

$\bar{Y}_{pjn}$    = 1, If product $p$ is assigned to unit $j$ in period $n$

$Y_{rn}^{R}$    = 1, If batch recipe $r$ is produced in period $n$

$Z_{f'j'fjn}$    = 1, If family $f'$, assigned to unit $j'$ in period $n$, is completed after starting family $f$, assigned to unit $j \neq j'$ in the same period $n$

# References

Jakeman CM (1994) Scheduling needs of the food processing industry. Food Res Int 27:117–120

Kopanos GM, Puigjaner L, Georgiadis MC (2010) Optimal production scheduling and lot-sizing in dairy plants: the yoghurt production line. Ind Eng Chem Res 49(2):701–718

Kopanos GM, Puigjaner L, Georgiadis MC (2012) Simultaneous production and logistics operations planning in semicontinuous food industries. OMEGA—Int J Manag Sci 40:634–650

Marchetti PA, Cerdá J (2009) A general resource-constrained scheduling framework for multistage batch facilities with sequence-dependent changeovers. Comput Chem Eng 33(4):871–886

Nakhla M (1995) Production control in the food processing industry. Int J Oper Prod Manag 15:73–88

Shah N (1998) Single- and multi-site planning and scheduling: current status and future challenge. In: Pekny JF, Blau GE (eds) Proceedings of the third international conference of the foundations of computer-aided process operations, AIChE symposium series 94, AIChE, New York, 1998, pp 75–90

Soman CA, Dp Van Donk, Gaalman GJC (2004) A basic period approach to the economic lot scheduling problem with shelf life considerations. Int J Prod Res 42:1677–1689

Van Dam P, Gaalman G, Sierksma G (1993) Scheduling of packaging lines in the process industry: an empirical investigation. Int J Prod Econ 30–31:579–589

# Chapter 7
# Simultaneous Optimization of Production and Logistics Operations in Semicontinuous Process Industries

## 7.1 Introduction

In the semicontinuous process industry, there is an ongoing trend toward an increased product variety and shorter replenishment cycle times. Hence, manufacturers seek a better coordination of production and distribution activities in order to avoid excessive inventories and improve customer's service. While traditionally minimizing production costs has been considered as the major objective, attention has shifted toward faster replenishment and improved logistical performance. Thus, finished product inventories are merely regarded as buffers between the manufacturing and the distribution stage of the supply chain. As a result, distribution costs have to be included in the overall objective function (Bilgen and Günther 2010).

In this chapter, the production and logistics operations planning in large-scale single- or multisite semicontinuous process industries is addressed (Kopanos et al. 2012). A new mixed discrete/continuous-time MIP model for the problem in question, based on the definition of families of products, is developed. A remarkable feature of the proposed approach is that in the production planning problem timing and sequencing decisions are taken for product families rather than for products. However, material balances are realized for every specific product, thus permitting the detailed optimization of production, inventory, and transportation costs. Sequence-dependent changeovers are also explicitly taken into account and optimized. Moreover, alternative transportation modes are considered for the delivery of final products from production sites to distribution centers. The efficiency and the applicability of the proposed approach are demonstrated by solving to optimality two industrial-size case studies, for an emerging real-life dairy industry, which is considered as a representative semicontinuous process industry.

© Springer Nature Switzerland AG 2019
G. M. Kopanos and L. Puigjaner, *Solving Large-Scale Production Scheduling and Planning in the Process Industries*,
https://doi.org/10.1007/978-3-030-01183-3_7

## 7.2　Problem Statement

In this chapter, the production and logistics operations planning problem in multisite multiproduct semicontinuous process industries is addressed. The basic features of the problem under consideration are summarized as follows:

(i)　A known planning horizon divided into a set of periods $n \in N$

(ii)　A set of production sites $s \in S$, and a set of distribution centers $d \in D$.

(iii)　A set of transportation trucks $l \in L$, which can transfer final products from production sites to distribution centers, $l \in L_{sd}$. Each transportation truck is characterized by a minimum and maximum capacity, $\varepsilon_t^{min}$ and $\varepsilon_t^{max}$, respectively.

(iv)　A set of processing units $j \in J$ which are installed on production site $s$, $j \in J_s$; with available processing time in period $n$ equal to $\omega_{sjn}$.

(v)　A set of products $p \in P$ with specific demand in period $n$, inventory costs $\xi_{spn}$, production rates $\rho_{psj}$ $j$, minimum processing runs $\mu_{psjn}^{min}$, processing costs $\theta_{psjn}$, and minimum storage time for processed products $\lambda_p$. $p_j$ is the subset of products that can be assigned to unit $j$, and $p_j$ is the subset of units that can produce product $p$.

(vi)　A set of batch recipes $r \in R$ (e.g., fermentation recipes) with minimum preparation time $\tau_r$, preparation cost $\chi_{srn}$, and minimum and maximum production capacity $\mu_{srn}^{min}$ and $\mu_{srn}^{max}$, respectively. The subset of products that come from batch recipe $r$ is denoted by $P_r$.

(vii)　A set of product families or simply *families* $f \in F$ wherein all products are grouped into; $F_j$ is the subset of families that can be assigned to unit $j$, and $P_f$ is the subset of units that can process family $f$, while the subset of products in family $f$ is denoted by $P_f$.

(viii)　A sequence-dependent changeover or simply *changeover* operation is required on each processing unit whenever the production is changed between two different families; the required changeover time is $\gamma_{ff'sj}$, while the changeover cost is $\phi_{ff'sjn}$.

(ix)　A sequence-independent setup operation, henceforth referred to as *setup* is required whenever product $p$ is assigned to a processing unit $j$; the setup time is $\delta_{psj}$.

We assume a non-preemptive operation mode, and no resource restrictions (e.g., manpower, steam, electricity, etc.).

The main key decision variables are

(i)　the optimal assignment of families and products to each processing unit in the production period, $Y_{fsjn}$ and $\bar{Y}_{psjn}$, respectively;

(ii)　the sequencing between families $f$ and $f'$ on each unit in every period, $X_{ff'sjn}$;

(iii)　the assignment of transportation trucks to processing sites—distribution center in each period $Z_{sdln}$ as well as the transportation load for each truck $\bar{U}_{sdln}$;

(iv)   the produced quantity for each product in each processing site at period $Q_{psjn}$ and the total produced amount of product $p$ per period $\bar{Q}_{psn}$; and, final;

(v)   the inventory profiles for each product at period $n$, $I_{spn}$.

The objective is to fully satisfy customer demand at minimum total cost, including production, changeover, and inventory and transportation costs.

## 7.3   Modeling Approach

As already discussed in Sect. 6.4, production planning in semicontinuous process industries typically deals with a large number of products with similar characteristics. This fact allows us to group products with similar characteristics into product families (families). In the proposed approach, products belong to the same family *if and only if*: (i) they come from the same batch recipe, and (ii) there is no need for changeover operations among them. Therefore, the production planning problem under question could be *partially* focused on families rather than on each product separately. More specifically, sequencing and timing decisions are taken for families and not for each separate product, as Fig. 7.1 illustrates. Obviously, the definition of families significantly reduces the size of the underlying mathematical model and, thus, the necessary computational burden without sacrificing any feasibility or optimality constraint.

A salient feature of the proposed mathematical formulation is the integration of three different modeling approaches (see Fig. 7.1). More specifically, we use: (i) a discrete-time approach for the calculation of inventories and transported quantities for products at the end of each period $n$ in the production and logistics operations planning level, (ii) a continuous-time approach for the sequencing of families in the scheduling level for families, and (iii) lot-sizing type capacity constraints in the short-term scheduling level for products. Further, it should be emphasized that the proposed modeling approach allows products that belong to the same family to have different: (i) processing rates (e.g., packing rates), (ii) operating costs, (iii) setup times, (iv) inventory costs, (v) transportation costs, and (vi) customer type.

## 7.4   Mathematical Formulation

In the proposed mathematical framework, constraints have been grouped according to the type of decision (assignment, timing, sequencing, etc.) upon which they are imposed on. For the sake of clarity of the model presented, we use uppercase Latin letters for decision variables and sets, and lowercase Greek letters for parameters.

**Fig. 7.1** Modeling approach

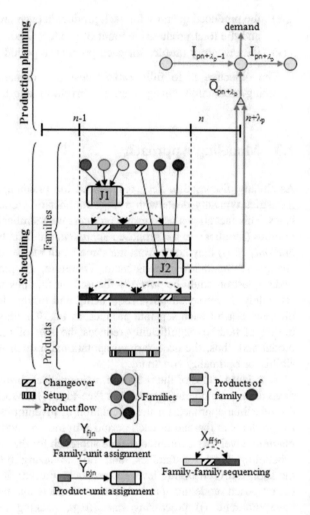

**Material Balance and Logistics Operations Constraints** The transportation of final products to customers (or distribution centers) is assumed to be done by three potential *transportation modes*: (a) transportation trucks owned by the customers, (b) transportation trucks owned by the industry, and (c) contracted transportation trucks from third-party logistics companies. Final products $p \in P_a$ whose final destination is the international market or big national supermarket customers are transported to their customers by transportation mode (a). The transportation of final products to the distribution centers owned by the enterprise can be performed by any of the other two transportation modes (b), and/or (c).

The total quantity of product $p$ produced in production plant $s$ in the period $n - \lambda_p$, which is ready to ship to customers in period $n$, is given by

$$\bar{Q}_{spn} = \sum_{j\in(J_s\cap J_p)} Q_{psjn-\lambda} \quad \forall s,p,n > \lambda_p \tag{7.1}$$

where $\lambda_p$ denotes the days that processed product $p$ should be kept in storage (e.g., for cooling or refrigeration purpose). $Q_{psjn}$ corresponds to the quantity of product $p$ processed in unit $j$ of production site $s$ during period $n$. It should be noticed that $\bar{Q}_{spn} = 0 \quad \forall s,p,n \leq \lambda_p$.

Constraint set (7.2) expresses the material balance of products $p \in P^a$, whose destination is the international market or big national supermarket clients.

$$I_{spn} = I_{spn-1} + \bar{Q}_{spn} - U_{spn}^a \quad \forall s,p \in P^a,n \tag{7.2}$$

where $U_{spn}^a$ denotes the quantity of product $p \in P^a$ transported from production site $s$ to the international market or big national supermarket clients by customer trucks, at period $n$, in order to fully meet the demand according to

$$\sum_s U_{spn}^a = \zeta_{pn}^a \quad \forall d,p \in P^a,n \tag{7.3}$$

$I_{spn}$ corresponds to the inventory level of product $p$ in production plant $s$ at time point $n$. Also, note that $I_{spn=0}$ reflects the initial inventory for product $p$ in production site $s$.

The multiperiod material balance constraints for products $p \notin P^a$ transported to company's distribution centers are given by

$$I_{spn} = I_{spn-1} + \bar{Q}_{spn} - \sum_{d\in D_s}\sum_{l\in L_{sd}} U_{sdlpn} \quad \forall s,p \notin P^a,n \tag{7.4}$$

where $U_{sdlpn}$ denotes the quantity of product $p$ transported from production site $s$ to distribution center $d$ by transportation truck $l$ at period $n$. Once final products reach distribution centers, they are stored for a day due to product quality purpose before sending them to final customers as follows:

$$\sum_{s\in S_d}\sum_{l\in L_{sd}} U_{sdlpn-1} = \zeta_{dnp} \quad \forall d,p \notin P^a,n > 1 \tag{7.5}$$

The total load for any transportation truck $l$ that transfers products from production facility $s$ to distribution center $d$ in period $n$ is calculated as follows:

$$\bar{U}_{sdln} = \sum_{p\notin P^a} U_{sdlpn} \quad \forall s,d \in D_s, l \in L_{sd},n \tag{7.6}$$

Hence, every truck $l$ has a specific minimum and maximum capacity ($\varepsilon_l^{\min}$, and $\varepsilon^{\max}$, respectively) as given by

$$\varepsilon_l^{\min} Z_{sdln} \leq \bar{U}_{sdln} \leq \varepsilon_l^{\max} Z_{sdln} \quad \forall s, d \in D_{s,l} \in L_{sd}, n \tag{7.7}$$

Binary variables $Z_{sdln}$ denote the use of truck $l$ for transporting products from production site $s$ to distribution center $d$ at period $n$. Any transportation truck $l$ can transfer products only between one production site $s$ and one distribution center $d$ during any period $n$, according to

$$\sum_{s \in S_l} \sum_{d \in (D_s \cap D_l)} Z_{sdln} \leq 1 \quad \forall l, n \tag{7.8}$$

**Product Lot-Sizing Constraints** Lower and upper bounds on the produced amounts of product $p$ are imposed by

$$\pi_{psjn}^{\min} \bar{Y}_{psjn} \leq Q_{psjn} \leq \pi_{psjn}^{\max} \bar{Y}_{psjn} \quad \forall p, s, j \in (J_s \cap J_p), n \tag{7.9}$$

Tighter maximum produced quantities for $p \in P^a$ can be estimated by

$$\pi_{psjn}^{\max} = \begin{cases} 0 & \text{if } \sum_{n' \geq n + \lambda_p}^{N} \zeta_{pn'}^a = 0 \\ (\omega_{jn} - \alpha_{jn} - \beta_{jn} - \sum_{r \in R_p} \tau_r)\rho_{psj} & \text{if } \sum_{n' \geq n + \lambda_p}^{N} \zeta_{pn'}^a \geq 0 \end{cases}$$

It should be noted that demands $\zeta_{pn'}^a$ must be met (i.e., full demand satisfaction). Similar expressions can be written for products $p \notin P^a$.

**Family Processing Time Definition** Because products that belong to the same family ($p \in P_f$) do not require changeover operations among them, sequencing and timing constraints should be solely imposed on families. However, it should be noted that setup times $\delta_{psj}$ might exist. In order to define sequencing and timing decisions for families, the definition of family processing time is introduced as follows:

$$T_{fsjn} = \sum_{p \in P_f} \left( \frac{Q_{psjn}}{\rho_{psj}} + \delta_{psj} \bar{Y}_{psjn} \right) \quad \forall f, s, j \in (J_f \cap J_s), n \tag{7.10}$$

Product processing rates, $\rho_{psj}$, are considered fixed as potential fluctuations may provoke quality problems (Soman et al. 2004).

**Family Allocation Constraints** A family $f$ is assigned to a processing unit $j$ of production site $s$ in period $n$ if at least one product $p \in P_f$, which belongs to this family, is processed in this unit during the same period:

$$Y_{fsjn} \geq \bar{Y}_{psjn} \quad \forall f, p \in P_f, s, j \in (J_f \cap J_s), n \tag{7.11}$$

Hence, constraint set (7.12) enforces the binary variables $Y_{fsjn}$ to zero when no products $p \in P_f$ are processed in unit $j$ at production site $s$ during period $n$.

$$Y_{fsjn} \leq \sum \bar{Y}_{psjn} \quad \forall f, s, j \in (J_f \cap J_s), n \tag{7.12}$$

**Family Sequencing and Timing Constraints** We introduce binary variables $X_{ff'sjn}$ to define the local precedence between two families $f$ and $f'$ in unit $j$, at production plant $s$ in period $n$. Constraints (7.13) and (7.14) state that, if a family $f$ is allocated to processing unit $j$ at production site $s$ in period $n$, (i.e., $Y_{fsjn} = 1$), then at most one family $f'$ is processed before and after it, respectively.

$$\sum_{f' \neq f, f' \in F_j} X_{f'fsjn} \leq Y_{fsjn} \quad \forall f, s, j \in (J_f \cap J_s), n \tag{7.13}$$

$$\sum_{f' \neq f, f' \in F_j} X_{ff'sjn} \leq Y_{fsjn} \quad \forall f, s, j \in (J_f \cap J_s), n \tag{7.14}$$

Obviously, the total number of active sequencing binary variables $X_{ff'sjn}$ plus the unit utilization binary variable $V_{sjn}$ should be equal to the total number of active allocation binary variables $Y_{fsjn}$ in a processing unit $j$ at production facility $s$ in period $n$, according to

$$\sum_{f \in F_j} \sum_{f' \neq f, f' \in F_j} X_{ff'sjn} + V_{sjn} = \sum_{f \in F_j} Y_{fsjn} \quad \forall s, j \in J_s, n \tag{7.15}$$

Constraint set (7.16) ensures that the processing unit $j$ in production site $s$ is used at period $n$, (i.e., $V_{sjn} = 1$) if at least one family $f$ is assigned to it over this period (i.e., $Y_{fsjn} = 1$). Note that no lower bound on the binary variable $V_{sjn}$ is necessary as far as a cost term (related to the unit utilization), is included into the objective function, thus enforcing $V_{sjn}$ to zero.

$$V_{sjn} \geq Y_{fsjn} \quad \forall f, s, j \in (J_f \cap J_s), n \tag{7.16}$$

The starting time of family $f'$, that directly follows another family $f$ on a processing line $j$ in production plant $s$ at period $n$, (i.e., $X_{ff'sjn} = 1$) should be greater than the completion time of family $f$, $C_{fsjn}$, plus the necessary changeover time $\gamma_{ff'sj}$ between those families:

$$C_{fsjn} + \gamma_{ff'sj} \leq C_{f'sjn} - T_{f'sjn} + M_{sjn}(1 - X_{ff'sjn}) \\ \forall f, f' \neq f, s, j \in (J_f \cap J_{f'} \cap J_s), n \tag{7.17}$$

Note that the big-M parameter $M_{sjn}$ can be set equal to $\omega_{sjn} - \beta_{sjn}$, where $\omega_{sjn}$ is the available production time horizon and $\beta_{sjn}$ corresponds to the daily plant shutdown time.

**Family Starting and Completion Time Bounds** Constraints (7.18) and (7.19) impose bounds on the starting and completion time of each family. More specifically, the starting time (i.e., $C_{fsjn} - T_{fsjn}$) has to be greater than the daily plant setup time, $\alpha_{sjn}$, plus the minimum batch time (fermentation process in the case of yogurt production) $\tau_r$ for preparing the recipe $r$, plus the changeover time $\gamma_{f'fsj}$ for changing the production to family $f$. An additional unit preparation time $o_{sjn}$ is also taken into account. This time stands for the additional preparation time of a processing unit $j$ due to potential maintenance or other technical reasons. Additionally, the release batch recipe time $\sigma_{srn}$ is also considered. In order to commence the production of a batch recipe $r$, all recipe ingredients need to be present. Otherwise, the production of the batch recipe $r$ will be postponed until the arrival of its missing substances.

$$
\begin{aligned}
C_{fsjn} - T_{fsjn} \geq & (\alpha_{sjn} + \max[o_{sjn}, \sigma_{srn}] + \tau_r) Y_{fsjn} \\
& + \sum_{f' \neq f\, f' \in F_j} \gamma_{f'fsj} X_{f'fsjn} \quad \forall s, f, r \in R_f, j \in (J_f \cap J_s), n
\end{aligned}
\tag{7.18}
$$

Hence, the completion time of each family should be smaller than the daily production time horizon $\omega_{sjn}$ minus the daily plant shutdown time $\beta_{sjn}$ as follows:

$$
C_{fsjn} \leq (\omega_{sjn} - \beta_{sjn}) Y_{fsjn} \quad \forall f, s, f \in (J_f \cap J_s), n
\tag{7.19}
$$

Production line shutdown is realized on a daily basis, as a typical production policy to guarantee the high quality of final products and to comply with hygienic standards.

**Batch Recipe Stage Constraints** Batch recipe stage (e.g., fermentation and pasteurization) constraints must be included into the mathematical model in order to ensure a feasible production plan. The cumulative produced quantity of products $p \in P_r$ should be greater than the minimum produced recipe amount in the batch recipe stages $\mu_{srn}^{min}$ and lower than the maximum production capacity $\mu_{srn}^{max}$:

$$
\mu_{srn}^{min} W_{srn} \leq \sum_{p \in P_r} \sum_{j \in (J_p \cap J_s)} Q_{psjn} \leq \mu_{srn}^{max} W_{srn} \quad \forall s, r, n
\tag{7.20}
$$

Constraint set (7.21) states that a batch recipe $r$ is produced in production facility $s$ at period $n$, (i.e., $W_{srn} = 1$), if at least one family $f \in F_r$ is processed on a processing unit $j$ in production site $s$ at the same period $n$ (i.e., $Y_{fsjn} = 1$).

$$
W_{srn} \geq \sum_{j \in (J_f \cap J_s)} Y_{fsjn} \quad \forall s, r, f \in F_r, n
\tag{7.21}
$$

**Tightening Constraints** In order to reduce the computational effort, constraint set (7.22) can further tighten the linear relaxation of the proposed mathematical model by imposing an upper bound on the total processing time for each processing line $j$ at each period $n$.

$$\sum_{f \in F_j} T_{fsjn} + \sum_{f \in F_j} \sum_{f' \neq f, f' \in F_j} \gamma_{ff'sj} X_{ff'sjn}$$
$$\leq (\omega_{sjn} - \alpha_{sjn} - \beta_{sjn} - \min_{r \in R_j}[\tau_r]) V_{sjn} \quad \forall s, j \in J_s, n \tag{7.22}$$

It should be noted that by incorporating constraint set (7.22) into the mathematical formulation, constraint set (7.16) could be omitted, thus further reducing the model size. Similarly, an upper bound on the family processing time can be defined as follows:

$$T_{fsjn} + \sum_{f' \neq f, f' \in F_j} \gamma_{ff'sj} X_{ff'sjn} \leq (\omega_{sjn} - \alpha_{sjn} - \beta_{sjn} - \tau_r) Y_{fsjn}$$
$$\forall f, r \in R_f, s, j \in (J_f \cap J_s), n \tag{7.23}$$

**Objective Function** The objective function to be minimized is the total cost including several factors such as: (i) inventory costs, (ii) operating costs, (iii) batch recipes preparation costs, (iv) unit utilization costs, (v) families changeover costs, and (vi) transportation costs, as follows:

$$\min \sum_s \sum_p \sum_n \xi_{spn} I_{spn} + \sum_s \sum_p \sum_{j \in (J_s \cap J_p)} \sum_n \frac{\theta_{psjn}}{\rho_{psj}} Q_{psjn}$$
$$+ \sum_s \sum_r \sum_n \chi_{srn} W_{srn} + \sum_s \sum_{j \in J_s} \sum_n v_{sjn} V_{sjn}$$
$$+ \sum_s \sum_f \sum_{f' \neq f} \sum_{j \in (J_s \cap J_f \cap J_{f'})} \sum_n \phi_{ff'sjn} X_{ff'sjn}$$
$$+ \sum_s \sum_{s \in D_s} \sum_{l \in L_{sd}} \sum_n (\psi_{sl} Z_{sdln} + v_{sdl} \bar{U}_{sdln}) \tag{7.24}$$

In a dairy plant, final yogurt products are kept at low temperatures, thus resulting in significant inventory cost (mainly due to high energy requirements), which should be considered in the optimization procedure. It should be also noted that the short shelf lives of yogurt products are indirectly taken into account through the inventory costs. Operating costs mainly include labor and energy costs plus costs due to material losses. The batch recipe (e.g., fermentation) cost account for all costs associated with the preparation of each batch recipe. The unit utilization cost basically stands for the shutdown cleaning operation cost, and the initial unit setup cost. Changeover costs correspond to cleaning and/or sterilization operations for switchover operations between families. Transportation costs include a fixed costs term for contracting the transportation vehicles and a variable costs term for the quantities transferred from sites to distribution centers. Note that products $p \in P^a$—whose final destination is the international market or big national supermarket customers—are transported by customers' trucks, and, therefore, there is no transportation cost for the industry. Finally, it should be mentioned that since full demand satisfaction is required, the minimization of total costs is identical to the maximization of total profit. The overall MIP model optimizes objective function (7.24) subject to constraints (7.1)–(7.23).

## 7.5   Case Studies

In this section, two industrial-size case studies are considered using the proposed MIP model. The first case (Case Study I) concerns the single-site production (already described in Sect. 6.6) and distribution planning of an emerging Greek dairy industry. Real data have been slightly modified due to confidentiality issues. The second case (Case Study II) considers the planning problem of multisite production and distribution, and is inspired by Case Study I.

At this point, it should be emphasized that in semicontinuous process plants, as well as in many other food processing industries, a natural sequence of products often exists (e.g., from the lower taste to the stronger or from the brighter color to the darker) thus the relative sequence of products within a family is known a priory. Therefore, when changing the production between two products of the same family, cleaning and sterilizing can be neglected. Hence, in such production plants not only the relative sequence of products belonging to the same product family may be known but also the relative sequence of families in each unit. In this case, different families are enumerated according to their relative position within the day. Table 7.1 illustrates the families' relative sequence inside a planning period for the case studies under consideration.

There are no initial inventories and setup times for products in both cases. Finally, all case studies have been solved to global optimality in a Dell Inspiron 1520 2.0 GHz with 2 GB RAM using CPLEX 11 under standard configurations via a GAMS 22.8 interface.

**Table 7.1** Families relative sequences in a planning period per unit

| Units | Families relative sequence |
|-------|----------------------------|
| J1 | F20 => F21 => F22 |
| J2 | F12 => F11 => F19 => F18 => 13 => 14 => F15 => F16 => F17 |
| J3 | F01 => F02 =>F03 => F05 => F04 = F08 = F09 => F10 => F06 => F07 |
| J4 | F08 => F09 => F10 => F06 => F23 |

### 7.5.1 Case Study I

The production and distribution network of the dairy industry under study consists of one production site and three distribution centers, as shown in Fig. 7.2. The production facility, situated in the city of Serres, has to fully satisfy the demand for: (i) products $p \notin P_a$ for the distribution centers, and (ii) products $p \in P_a$ for international customers and big local supermarket clients. The plant operates with 4 packing lines (J1–J4). The 93 final products (P01–P93) are grouped into 23 families (F01–F23). The dairy industry owns a pair of transportation trucks (OWN-1, and OWN-2) with $\psi_{sl} = 50 \, €$ and minimum and maximum load capacity, $\varepsilon_l^{min} = 1000 \, kg$ and $\varepsilon_l^{max} = 6000 \, kg$, respectively. These trucks can supply the distribution centers situated in Thessaloniki and Xanthi. Ten third-party logistics trucks (3PLT-1 to 3PLT-10) with $\psi_{sl} = 700 \, €$ and $\varepsilon_l^{min} = 1000 \, kg$ and $\varepsilon_l^{max} = 12,000 \, kg$ are also available. 3PLT trucks can supply the distribution centers in Thessaloniki and Athens. The remaining data are not provided due to confidentiality issues.

**Fig. 7.2** Case study I: production site and distribution centers locations

The resulting mathematical model consists of 9639 constraints, 2160 binary variables, and 15,462 continuous variables. The optimal solution, corresponding to a total cost of 436,167 €, was reached in just 7.7 CPUs after exploring 610 nodes in the branch-and-bound tree.

Figure 7.3 presents the total cost breakdown as well as the contribution of each cost term in the total cost. Inventory and transportation costs stand for the 61.0% of the total cost while production costs (i.e., operating, recipe, unit utilization, and changeovers costs) represent the 39.0% of the total cost. The profiles of the total produced quantities, inventories, and transported quantities for each planning period are shown in Fig. 7.4.

The production site operates from n0 to n5 period. Also, note that due to high demand requirements the production facility operates in period n0, which is an overtime period (i.e., higher operating costs). The highest total production is observed in period n5, with 83,463 kg of production. In period n6, a very high inventory level of 81,165 kg is detected. The transportation schedule is realized

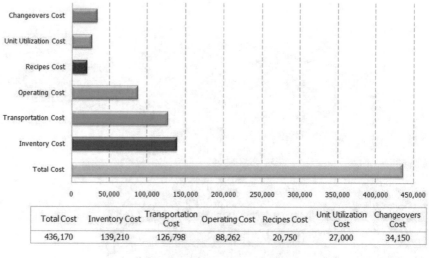

| Total Cost | Inventory Cost | Transportation Cost | Operating Cost | Recipes Cost | Unit Utilization Cost | Changeovers Cost |
|---|---|---|---|---|---|---|
| 436,170 | 139,210 | 126,798 | 88,262 | 20,750 | 27,000 | 34,150 |

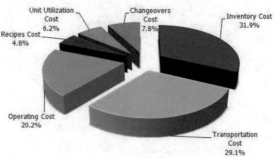

**Fig. 7.3** Case study I: total cost breakdown (€) and cost terms contribution

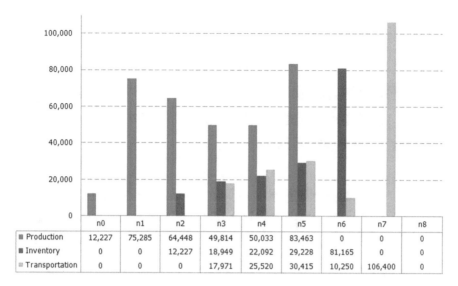

**Fig. 7.4** Case study I: production, inventory, and transportation profiles per period (kg)

from n3 to n7 period. The peak of transportation quantity is observed in period n7 where a total of 106,400 kg is transferred from the production site to the distribution centers.

Figure 7.5 illustrates the detailed production plan for families. The sequences between families can be found in Table 7.1. The optimal transportation plan is given in Table 7.2. A total number of 11 trucks are occupied in period n7, wherein the peak of logistics operations is observed as Fig. 7.4 illustrates.

Moreover, the proposed MIP formulation provides us with the detailed transportation plan for each product (i.e., assignment of product to truck, assignment of the truck to the distribution center, and quantity of product transported by each truck). An example is presented in Table 7.3 where the detailed product transportation plan in period's n3 and n4 is shown.

## 7.5.2  Case Study II

This case is concerned with the multisite production and logistics operations planning problem. The production and distribution network under consideration consists of two production sites (situated in Serres, and Karditsa) and five distribution centers, as shown in Fig. 7.6. Processing units J1–J4 are installed in the production plant situated in Serres while processing units J1–J3 are installed on the production site of Karditsa. The production plants have to fully meet the demand for all products. The production site in Serres owns a pair of transportation trucks (OWN-1, and OWN-2) with $\psi_{s,l} = 50 \, €$ and minimum and maximum load

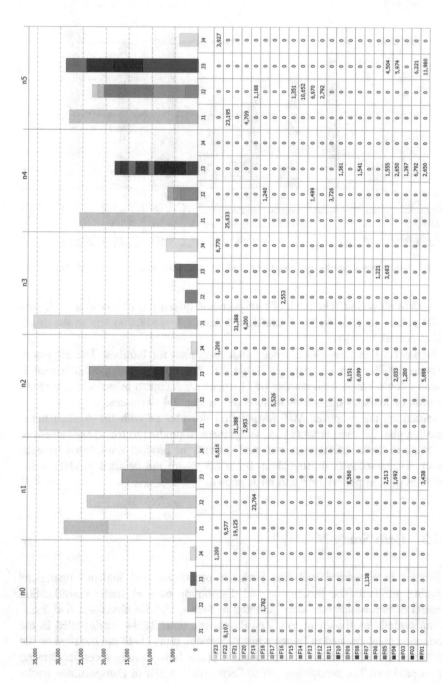

**Fig. 7.5** Case study I: production plan for families (kg)

**Table 7.2**  Case study I: transportation plan (kg)

| Truck1 | Distr.Center | n0 | n1 | n2 | n3 | n4 | n5 | n6 | n7 | n8 |
|--------|--------------|----|----|----|----|----|----|----|----|----|
| OWN-1  | Thessaloniki | 0 | 0 | 0 | 6000 | 6000 | 0 | 5250 | 6000 | 0 |
|        | Xanthi       | 0 | 0 | 0 | 0 | 0 | 4261 | 0 | 0 | 0 |
| OWN-2  | Thessaloniki | 0 | 0 | 0 | 4506 | 0 | 0 | 0 | 0 | 0 |
|        | Xanthi       | 0 | 0 | 0 | 0 | 4151 | 4261 | 0 | 5924 | 0 |
| 3PLT-1 | Athens       | 0 | 0 | 0 | 7465 | 8918 | 0 | 5000 | 10,795 | 0 |
|        | Thessaloniki | 0 | 0 | 0 | 0 | 0 | 9766 | 0 | 0 | 0 |
| 3PLT-2 | Athens       | 0 | 0 | 0 | 0 | 0 | 0 | 0 | 0 | 0 |
|        | Thessaloniki | 0 | 0 | 0 | 0 | 6450 | 5699 | 0 | 10,795 | 0 |
| 3PLT-3 | Athens       | 0 | 0 | 0 | 0 | 0 | 5699 | 0 | 10,795 | 0 |
|        | Thessaloniki | 0 | 0 | 0 | 0 | 0 | 0 | 0 | 0 | 0 |
| 3PLT-4 | Athens       | 0 | 0 | 0 | 0 | 0 | 0 | 0 | 0 | 0 |
|        | Thessaloniki | 0 | 0 | 0 | 0 | 0 | 0 | 0 | 10,795 | 0 |
| 3PLT-5 | Athens       | 0 | 0 | 0 | 0 | 0 | 0 | 0 | 0 | 0 |
|        | Thessaloniki | 0 | 0 | 0 | 0 | 0 | 0 | 0 | 10,795 | 0 |
| 3PLT-6 | Athens       | 0 | 0 | 0 | 0 | 0 | 0 | 0 | 0 | 0 |
|        | Thessaloniki | 0 | 0 | 0 | 0 | 0 | 0 | 0 | 10,795 | 0 |
| 3PLT-7 | Athens       | 0 | 0 | 0 | 0 | 0 | 0 | 0 | 9901 | 0 |
|        | Thessaloniki | 0 | 0 | 0 | 0 | 0 | 0 | 0 | 0 | 0 |
| 3PLT-8 | Athens       | 0 | 0 | 0 | 0 | 0 | 0 | 0 | 9901 | 0 |
|        | Thessaloniki | 0 | 0 | 0 | 0 | 0 | 0 | 0 | 0 | 0 |
| 3PLT-9 | Athens       | 0 | 0 | 0 | 0 | 0 | 0 | 0 | 0 | 0 |
|        | Thessaloniki | 0 | 0 | 0 | 0 | 0 | 0 | 0 | 9901 | 0 |

capacity, $\varepsilon_l^{min} = 1000\,kg$ and $\varepsilon_l^{max} = 12,000\,kg$, respectively. These trucks can supply distribution centers located at Thessaloniki and Xanthi. Ten third-party logistics trucks (3PLT-1 to 3PLT-10) with $\psi_{s,l} = 700$ € and $\varepsilon_l^{min} = 1000\,kg$ and $\psi_{s,l} = 12,000$ € are also available in the production plant of Serres. The distribution centers of Thessaloniki, Athens, and Ioannina can be supplied from the production site in Serres by 3PLT trucks. The production facility located to Karditsa owns a pair of transportation trucks (OWN-3, and OWN-4) with $\psi_{sl} = 50$ € and minimum and maximum load capacity, $\varepsilon_l^{min} = 1000\,kg$ and $\varepsilon_l^{max} = 6,000\,kg$, respectively. These trucks can supply distribution centers situated in Ioannina, and Patras. In addition, six third-party logistics trucks (3PLT-11 to 3PLT-16) with $\psi_{s,l} = 600$ € and $\varepsilon_l^{min} = 1000\,kg$ and $\varepsilon_l^{max} = 6000\,kg$ and $\varepsilon_l^{max} = 12,000\,kg$ are available in the production facility of Karditsa. The distribution centers of Thessaloniki, Athens, and Ioannina can be supplied from the production site in Karditsa by 3PLT trucks. Notice that the production facility in Serres cannot supply the distribution center of Patras, and the production plant in Karditsa cannot supply the distribution center of Xanthi. The remaining data are not provided due to confidentiality issues.

**Table 7.3** Case Study I: Detailed transportation plan for period n3 and n4 (kg)

| Distr. center | Truck | Product | n3 | Distr. center | Truck | Product | n4 |
|---|---|---|---|---|---|---|---|
| Athens | 3PLT-1 | P01 | 2538 | Athens | 3PLT-1 | P28 | 1850 |
| | | P12 | 1152 | | | P29 | 1850 |
| | | P17 | 1703 | | | P34 | 1325 |
| | | P90 | 438 | | | P35 | 1325 |
| | | P91 | 864 | | | P02 | 1308 |
| | | P92 | 578 | | | P13 | 1098 |
| | | P93 | 192 | | | P64 | 164 |
| | | Total | 7465 | | | Total | 9918 |
| Thesaloniki | OWN-1 | P84 | 1500 | Thesaloniki | 3PLT-2 | P34 | 3435 |
| | | P90 | 909 | | | P28 | 1200 |
| | | P91 | 1836 | | | P29 | 1200 |
| | | P92 | 923 | | | P13 | 616 |
| | | P93 | 832 | | | Total | 6450 |
| | | Total | 6000 | | | | |
| | | | | Thesaloniki | OWN-1 | | |
| Thesaloniki | OWN-2 | P01 | 900 | | | | |
| | | P12 | 540 | | | | |
| | | P17 | 810 | | | | |
| | | P88 | 1013 | | | | |
| | | P90 | 1244 | | | | |
| | | Total | 4506 | | | | |
| | | | | Xanthi | OWN-2 | P02 | 1548 |
| | | | | | | P04 | 864 |
| | | | | | | P01 | 852 |
| | | | | | | P10 | 374 |
| | | | | | | P12 | 175 |
| | | | | | | P05 | 149 |
| | | | | | | P13 | 144 |
| | | | | | | P64 | 45 |
| | | | | | | Total | 4151 |

The resulting mathematical model consists of 19,371 constraints, 4070 binary variables, and 35,614 continuous variables. The optimal solution corresponds to 520,047 € of total cost, and it was obtained in 379.2 CPUs after exploring 1837 nodes in the branch-and-bound tree. The profile of total produced quantities, inventories, and transported quantities for each planning period are illustrated in Fig. 7.7. The production plants work from n0 to n5 period, where period n0 is an overtime period (i.e., higher operating costs). The highest production is observed in period n5, with 129,723 kg of total production. Generally speaking, inventory levels are maintained low throughout the planning horizon, with an exception in

**Fig. 7.6** Case study II: production sites and distribution centers locations

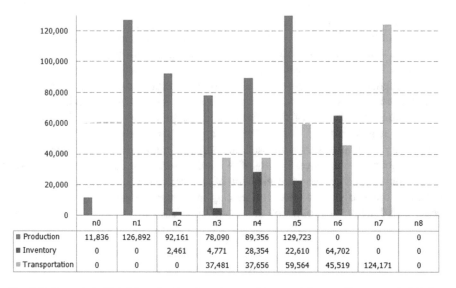

| | n0 | n1 | n2 | n3 | n4 | n5 | n6 | n7 | n8 |
|---|---|---|---|---|---|---|---|---|---|
| ■ Production | 11,836 | 126,892 | 92,161 | 78,090 | 89,356 | 129,723 | 0 | 0 | 0 |
| ■ Inventory | 0 | 0 | 2,461 | 4,771 | 28,354 | 22,610 | 64,702 | 0 | 0 |
| ▨ Transportation | 0 | 0 | 0 | 37,481 | 37,656 | 59,564 | 45,519 | 124,171 | 0 |

**Fig. 7.7** Case study II: total production, inventory, and transportation profiles per period (kg)

period n6 where a relatively high inventory level of 64,702 kg is detected. Transportation operations from production facilities to distribution centers are realized from n3 to n7 period. In period n7, a total of 124,171 kg of products is transferred from the production sites to the distribution centers

Figure 7.8 illustrates the total, and per production facility, cost breakdown as well as the contribution of each cost term on the total cost. Inventories cost represent 14.5% of the total cost while transportation costs stand for the 35.7% of the total cost. Production costs (i.e., operating, recipe, unit utilization, and changeovers costs) represent 49.8% of the total cost. Also note that 62.5% of the total inventory cost and 57.5% of the total transportation cost occurred in the production site of Serres.

Figure 7.9 presents the detailed production plan for the production plant in Serres, and Fig. 7.10 shows the detailed production plan for the production facility of Karditsa. The sequences between families are predetermined (see Table 7.1). The total amount of products transported from production facilities to distribution centers by transportation trucks is given in Table 7.4. A total of 14 trucks are needed in period n7, wherein the peak of logistics operations is observed as illustrated in Fig. 7.7. Finally, the proposed MIP model also generates the detailed transportation plan for each product (i.e., assignment of product to truck, assignment of truck to distribution center, and quantity of product transported by each truck).

## 7.6   Concluding Remarks

In this chapter, we have developed a MIP formulation, based on the definition of families of products, for the simultaneous optimization of single or multisite production and logistics operations in semicontinuous process industries. Two industrial-size case studies for a real-life dairy industry have been solved to optimality in order to shed light on the special features of the suggested MIP model. It should be emphasized that while production timing and sequencing decisions are taken for families (rather than for products), material balances are realized for each specific product, thus permitting the detailed optimization of production, inventory, and transportation costs. Additionally, alternative transportation modes are considered for the delivery of final products from production sites to distribution centers, a reality that most of the current approaches totally neglect. Despite the complexity of the problems addressed, the proposed approach appears a remarkable computational performance. Finally, it is worth mentioning that the proposed MIP model aims at being the core element of a computer-aided advanced planning system in order to facilitate decision making in relevant industrial environments by better coordination of production and distribution activities.

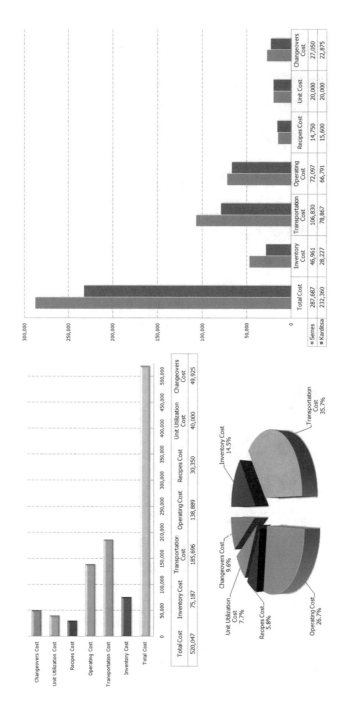

| | Inventory Cost | Transportation Cost | Operating Cost | Recipes Cost | Unit Utilization Cost | Changeovers Cost | Total Cost |
|---|---|---|---|---|---|---|---|

**Fig. 7.8** Case study II: total cost breakdown (€) and cost terms contribution

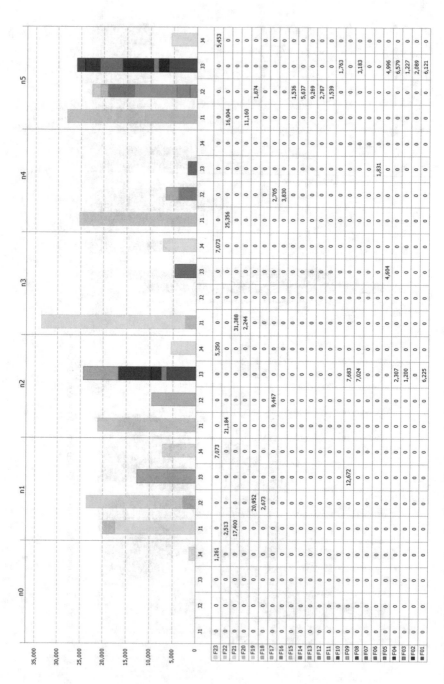

**Fig. 7.9** Case study II: production plan for families in production facility in Serres (kg)

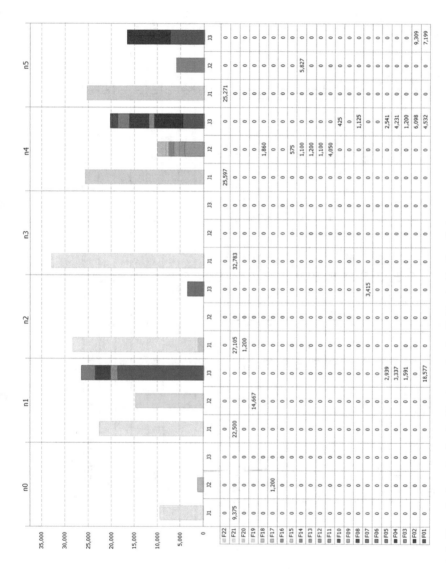

**Fig. 7.10** Case study II: production plan for families in production facility in Karditsa (kg)

**Table 7.4** Case study II: transportation plan (kg)

| Site | Truck | Distr. center | n0 | n1 | n2 | n3 | n4 | n5 | n6 | n7 | n8 |
|---|---|---|---|---|---|---|---|---|---|---|---|
| Serres | OWN-1 | Thessaloniki | 0 | 0 | 0 | 6000 | 0 | 0 | 6000 | 0 | 0 |
| | | Xanthi | 0 | 0 | 0 | 0 | 4151 | 3241 | 0 | 5924 | 0 |
| | OWN-2 | Thessaloniki | 0 | 0 | 0 | 2256 | 6000 | 0 | 6000 | 6000 | 0 |
| | | Xanthi | 0 | 0 | 0 | 0 | 0 | 6000 | 0 | 0 | 0 |
| | | Athens | 0 | 0 | 0 | 2590 | 11,143 | 0 | 0 | 0 | 0 |
| | 3PLT-1 | Thessaloniki | 0 | 0 | 0 | 0 | 0 | 3475 | 12,000 | 11,836 | 0 |
| | | Ioannina | 0 | 0 | 0 | 0 | 0 | 0 | 0 | 0 | 0 |
| | | Athens | 0 | 0 | 0 | 0 | 0 | 0 | 0 | 0 | 0 |
| | 3PIT-2 | Thessaloniki | 0 | 0 | 0 | 0 | 6450 | 12,000 | 0 | 11,932 | 0 |
| | | Ioannina | 0 | 0 | 0 | 0 | 0 | 0 | 0 | 0 | 0 |
| | | Athens | 0 | 0 | 0 | 0 | 0 | 7124 | 0 | 0 | 0 |
| | 3PLT-3 | Thessaloniki | 0 | 0 | 0 | 0 | 0 | 0 | 0 | 0 | 0 |
| | | Ioannina | 0 | 0 | 0 | 0 | 0 | 0 | 0 | 6549 | 0 |
| | | Athens | 0 | 0 | 0 | 0 | 0 | 0 | 0 | 0 | 0 |
| | 3PLT-4 | Thessaloniki | 0 | 0 | 0 | 0 | 0 | 0 | 0 | 10,697 | 0 |
| | | Ioannina | 0 | 0 | 0 | 0 | 0 | 4830 | 0 | 0 | 0 |
| | | Athens | 0 | 0 | 0 | 0 | 0 | 0 | 0 | 10,582 | 0 |
| | 3PLT-5 | Thessaloniki | 0 | 0 | 0 | 0 | 0 | 0 | 0 | 0 | 0 |
| | | Ioannina | 0 | 0 | 0 | 0 | 0 | 0 | 0 | 0 | 0 |
| | | Athens | 0 | 0 | 0 | 0 | 0 | 0 | 0 | 0 | 0 |
| | 3PLT-6 | Thessaloniki | 0 | 0 | 0 | 0 | 0 | 0 | 0 | 6812 | 0 |
| | | Ioannina | 0 | 0 | 0 | 0 | 0 | 0 | 0 | 0 | 0 |

(continued)

**Table 7.4** (continued)

| Site | Truck | Distr. center | n0 | n1 | n2 | n3 | n4 | n5 | n6 | n7 | n8 |
|------|-------|---------------|----|----|----|----|----|----|----|----|----|
| Karditsa | OWN-3 | Ioannina | 0 | 0 | 0 | 0 | 0 | 6000 | 0 | 5652 | 0 |
| | | Patras | 0 | 0 | 0 | 5189 | 3906 | 0 | 5769 | 0 | 0 |
| | OWN-4 | Ioannina | 0 | 0 | 0 | 6000 | 0 | 6000 | 0 | 5569 | 0 |
| | | Patras | 0 | 0 | 0 | 6000 | 6000 | 0 | 1000 | 0 | 0 |
| | | Athens | 0 | 0 | 0 | 0 | 0 | 0 | 11,500 | 12,000 | 0 |
| | 3PLT-11 | Thessaloniki | 0 | 0 | 0 | 2250 | 0 | 0 | 0 | 0 | 0 |
| | | Ioannina | 0 | 0 | 0 | 0 | 0 | 10,894 | 0 | 0 | 0 |
| | | Athens | 0 | 0 | 0 | 6741 | 0 | 0 | 0 | 0 | 0 |
| | 3PLT-12 | Thessaloniki | 0 | 0 | 0 | 0 | 0 | 0 | 2250 | 12,000 | 0 |
| | | Ioannina | 0 | 0 | 0 | 0 | 0 | 0 | 0 | 0 | 0 |
| | | Athens | 0 | 0 | 0 | 0 | 0 | 0 | 1000 | 0 | 0 |
| | 3PLT-13 | Thessaloniki | 0 | 0 | 0 | 0 | 0 | 0 | 0 | 6567 | 0 |
| | | Ioannina | 0 | 0 | 0 | 6454 | 0 | 0 | 0 | 0 | 0 |
| | | Athens | 0 | 0 | 0 | 0 | 0 | 0 | 0 | 12,000 | 0 |
| | 3PLT-14 | Thessaloniki | 0 | 0 | 0 | 0 | 0 | 0 | 0 | 0 | 0 |
| | | Ioannina | 0 | 0 | 0 | 0 | 0 | 0 | 0 | 0 | 0 |

## 7.7   Nomenclature

### Indices/Sets

$d \in D$    Distribution centers
$f, f' \in F$    Product families (families)
$j, j' \in J$    Processing unit types (units)
$k \in K$    Renewable resources
$n \in N$    Planning time periods
$p \in P$    Products
$r \in R$    Batch recipes (recipes)
$s \in S$    Production sites

### Subsets

$D_l$    Distribution centers $d$ that can be supplied by truck $l$
$D_s$    Distribution centers $d$ that can be supplied by production site $s$
$F_j$    Families $f$ that can be processed on unit $j$
$F_r$    Families $f$ that have the same recipe type origin $r$
$J_f$    Available units $j$ to process family $f$
$J_p$    Units $j$ that can process product $p$
$J_s$    Processing units $j$ that that are installed on production site $s$
$L_{sd}$    Transportation trucks $l$ that can transfer products from production site $s$ to distribution center $d$
$P^a$    Products $p$ that are destined for international customers or big national supermarket clients, which have their own trucks
$P_f$    Products $p$ that belong to the same family $f$
$P_r$    Products $p$ that have the same recipe origin $r$
$R_r$    Recipe origin $r$ for family $f$
$R_j$    Recipes $r$ that can be processed in unit $j$
$R_p$    Product $p$ that comes from recipe $r$
$S_d$    Production sites $s$ that can supply distribution center $d$
$S_l$    Production sites $s$ that can use transportation truck $l$

### Parameters

$\alpha_{sjn}$    Daily opening setup time for every unit $j$ of production site $s$ in period $n$; accounts for the pasteurization and homogenization stages
$\beta_{jn}$    Daily shutdown time for every unit $j$ in period $n$ (e.g., cleaning of yogurt production line for hygienic and quality reasons)
$\gamma_{ff'sj}$    Changeover time between family $f$ and family $f'$ on unit $j$ of production site $s$; accounts for cleaning and sterilizing operations
$\delta_{spj}$    Setup time for product $p$ on unit $j$ of production site $s$
$\varepsilon_l^{max}$    Maximum capacity of transportation truck l
$\varepsilon_l^{min}$    Minimum capacity of transportation truck $l$
$\zeta_{dpn}$    Demand for product $p \notin P^a$ of customers supplied by distribution center $d$ at time $n$

$\zeta_{pn}^{a}$      Demand for product $p \in P^{a}$ at time $n$

$\theta_{psjn}$      Variable operating cost for product $p$ on processing unit $j$ of production site $s$ in period $n$; includes labor and utilities costs

$\lambda_{p}$      Minimum cooling storage time for processed products (in periods n)

$M_{sjn}$      A big number

$\mu_{srn}^{max}$      Maximum production capacity of batch recipe $r$ in production site $s$ in period $n$

$\mu_{rn}^{min}$      Minimum produced quantity of batch recipe $r$ in production site $s$ in period $n$; accounts for pasteurization and fermentation tanks capacity restrictions

$v_{sjn}$      Fixed cost for utilizing unit $j$ of production site $s$ in period $n$

$\xi_{spn}$      Inventory cost for product $p$ in production site $s$ in period $n$

$o_{jn}$      Additional unit preparation time for processing unit $j$ of production site $s$ in periods $n$

$\pi_{psjn}^{max}$      Maximum production runs for product $p$ on unit $j$ of production site $s$ in period $n$

$\pi_{psjn}^{min}$      Minimum production runs for product $p$ on unit $j$ of production site $s$ in period $n$

$\rho_{psj}$      Processing rate for product $p$ on unit $j \in J_{p}$ of production site $s$

$\sigma_{srn}$      Release time for recipe $r$ in period $n$ of production site s

$\tau_{r}$      Minimum time for preparing recipe $r$; (e.g., for producing stirred yogurt products stands for the minimum fermentation time, while for set yogurt products reflects the minimum cooling time before the packing stage)

$\upsilon_{sdl}$      Variable cost for transferring products from production site $s$ to distribution center $d$ by truck $l$

$\phi_{ff'sjn}$      Changeover cost between family $f$ and family $f'$ in unit $j$ of production site $s$ in period $n$; accounts for cleaning and sterilizing operations

$\chi_{srn}$      Cost for producing batch recipe $r$ in production site $s$ in period $n$

$\psi_{sl}$      Fix cost for contracting transportation truck $l$ to carry products from production site $s$

$\omega_{jn}$      Physical available processing time in unit $j$ at period $n$

**Continuous Variables**

$C_{fsjn}$      Completion time for family $f$ in unit $j$ of production site $s$ in period $n$

$I_{spn}$      Inventory of product $p$ in production site $s$ at time $n$

$Q_{psjn}$      Produced amount of product $p$ in unit $j$ of production site $s$ in period $n$

$\bar{Q}_{psn}$      Total produced amount of product $p$ in production site $s$ in period $n$

$T_{fsjn}$      Processing time for family $f$ in unit $j$ of production site $s$ in period $n$

$U_{sdlpn}$      Quantity of product $p \in P^{a}$ transported from production site $s$ to distribution center $d$ by truck $l$ in period $n$

$\bar{U}_{sdln}$      Total transported quantity from production site $s$ to distribution center $d$ by truck $l$ in period $n$

$U_{spn}^{a}$      Quantity of product $p \in P^{a}$ transported from production site $s$ to international market or big national supermarket clients by customer trucks in period $n$

**Binary Variables**

$V_{sjn}$   = 1, If unit $j$ of production site $s$ is used in period $n$

$W_{srn}$   = 1, If batch recipe $r$ is produced in production site $s$ in period $n$

$X_{ff'sjn}$   = 1, If family $f'$ is processed exactly after family $f$, when both are assigned to the same unit $j$ of production site $s$ in the period $n$

$Y_{fsjn}$   = 1, If family $f$ is assigned to unit $j$ of production site $s$ in period $n$

$\bar{Y}_{psjn}$   = 1, If product $p$ is assigned to unit $j$ of production site $s$ in period $n$

$Z_{sdln}$   = 1, If transportation truck $l$ transfers material from production facility $s$ to distribution center $d$ in period $n$

# References

Bilgen B, Günther HO (2010) Integrated production and distribution planning in the fast moving consumer goods industry: a block planning application. OR Spectrum 32:927–955

Kopanos GM, Puigjaner L, Georgiadis MC (2012) Simultaneous production and logistics operations planning in semicontinuous food industries. OMEGA—Int J Manag Sci 40:634–650

Soman CA, Dp Van Donk, Gaalman GJC (2004) A basic period approach to the economic lot scheduling problem with shelf life considerations. International Journal of Production Research 42:1677–1689

# Part IV
# Batch Processes

# Chapter 8
# Production Scheduling in Large-Scale Multistage Batch Process Industries

## 8.1 Introduction

Nowadays, it is widely recognized that the current gap between practice and theory in the area of short-term scheduling needs to be bridged, as clearly remarked in Méndez et al. (2006) and Ruiz et al. (2008). New academic developments are mostly tested on relatively small problems whereas current real-world industrial applications consist of hundreds of batches, numerous multiple units available for each task and a long sequence of processing stages. Additionally, there exist a wide range of operational constraints that should be taken into account in order to guarantee the feasibility of the proposed schedule. Most industrial problems are very hard constrained, thus optimization solvers have to find optimal or near-optimal solutions in a huge search space with a relatively small feasible region. This fact may result in huge computational requirements that often do not allow finding even good feasible solutions, which is definitely not suitable for industrial environments.

Since most industrial scheduling problems are highly combinatorial and complex decision-making processes, they rarely can be solved to optimality within a reasonable computational time. In addition, the computational effort to find a good solution tends to be as important as the scheduling problem itself; since industry demands solutions that are both optimal, or at least close-optimal, and quick to be reached.

In this chapter, an efficient systematic iterative MIP-based solution strategy for solving real-world scheduling problems in multiproduct multistage batch plants is presented. A novel precedence-based concept has been also developed here. The proposed solution strategy consists of a constructive step, wherein a feasible and initial solution is rapidly generated by following an iterative insertion procedure, and an improvement step, wherein the initial solution is systematically enhanced by implementing iteratively several rescheduling techniques based on the mathematical model. A salient feature of our approach is that the scheduler can maintain the

© Springer Nature Switzerland AG 2019
G. M. Kopanos and L. Puigjaner, *Solving Large-Scale Production Scheduling and Planning in the Process Industries*,
https://doi.org/10.1007/978-3-030-01183-3_8

number of decisions at a reasonable level thus reducing appropriately the search space. This usually results in manageable model sizes that often favor a more stable and predictable optimization model behavior.

## 8.2  Problem Statement

The problem under consideration is concerned with industrial-scale multiproduct, multistage batch processes with the following features:

(i)   A set of product orders $i \in I$ should be processed by following a predefined sequence of processing stages $s \in S$ with, in general, unrelated processing units $j \in J$ working in parallel.

(ii)  Each product order $i$ comprises of a single batch that must follow a set of processing stages $s \in S_i$.

(iii) Some products $i$ may skip certain processing stages $s \notin S_i$, since different production recipes are considered.

(iv)  A product order $i$ can be processed in a specific subset of units $j \in J_i$. Similarly, a processing stage $s$ can be processed in a specific subset of units $j \in J_s$.

(v)   Transition times between consecutive product orders involve two terms. The first depends on both the unit and the order being processed $\pi_{ij}$ while the second also varies with the order previously manufactured in that unit $(\gamma_{ii'j})$. Transition times must be explicitly taken into account in the schedule generation process since they are usually of the same order of magnitude or even larger than the processing times. Consequently, they become a very critical feature when scheduling real-world batch processes such as pharmaceuticals, chemicals, food, etc.

(vi)  Model parameters like order due dates $(\delta_i)$, processing times $(\tau_{isj})$, unit-dependent setup times $(\pi_{ij})$, sequence-dependent setup (or simply changeover) times $(\gamma_{ii'j})$ and costs $(\xi_{ii'j})$, order release times $(oi)$, unit available times $(\varepsilon_j)$, and operating cost $(\psi)$ are all deterministic.

(vii) Once the processing of an order in a given stage is started, it should be carried out until completion without interruption (non-preemptive mode).

(viii) Mixing or splitting of product orders is not allowed.

The key decision variables are as follows:

(i)   the allocation of products $i$ to units $j \in J_i$ per stage, $Y_{isj}$;

(ii)  the relative sequence for any pair of products $i$, $i'$ in unit $j \in (J_i \cap J_{i'})$, $X_{ii'j}$;

(iii) the completion time of products $i$ in processing stage $s \in s_i$, $C_{is}$.

Alternative objective functions can be considered, such as the minimization of makespan, total weighted lateness or total operating and changeovers cost.

## 8.3  Mathematical Formulations

In this section, two batch-oriented mathematical models are presented for solving scheduling problems in multiproduct multistage batch plants (Kopanos et al. 2010). Both models are based on a continuous-time domain and utilize sequencing variables. The first model is based on the general (global) precedence sequencing concept, and the latter one is based on the unit-specific general precedence sequencing concept which has been developed as a part of this book (see Appendix B).

Global precedence formulations result in models with small model size and they are computationally faster on average. However, a drawback of these models is that they cannot optimize objectives containing changeover issues (e.g., minimization of changeover costs). For this reason, a unit-specific general precedence model, for scheduling multiproduct multistage batch plants, able to cope with a wide variety of objective functions, is also presented in the context of a more general mathematical formulation.

It is worth noticing that the MIP models, presented in this work, are not claimed to be either the fastest or the tightest. However, for the sake of clarity of the presentation of the proposed MIP-based solution strategy, the MIP models adopted were entirely developed along this work rather than using readily available models from the literature. Otherwise, other mathematical formulations found in the literature could be used as core MIP models in the proposed solution strategy. The description of the mathematical frameworks used in this work follows.

### 8.3.1  A General Precedence Multistage Scheduling Framework

The problem under study can be formulated by the following sets of constraints using the general precedence notion:

$$\sum_{j \in (J_i \cap J_s)} Y_{isj} = 1 \quad \forall i \in I^{\text{in}},\ s \in S_i \tag{8.1}$$

$$C_{is} \geq \sum_{j \in (J_i \cap J_s)} (\max[\varepsilon_j, o_i] + \pi_{ij} + \tau_{isj}) Y_{isj} \quad \forall i \in I^{\text{in}},\ s \in S_i : s = 1 \tag{8.2}$$

$$C_{is} - \sum_{j \in (J_i \cap J_s)} (\pi_{ij} + \tau_{isj}) Y_{isj} = C_{is-1} + W_{is-1} + \mu_{s-1s}$$
$$\forall i \in I^{\text{in}},\ s \in S_i : s > 1 \tag{8.3}$$

$$C_{is} + \gamma_{ii'j} \leq C_{i's} - \pi_{i'j} - \tau_{i'sj} + M(1 - X_{ii'j}) + M(2 - Y_{isj} - Y_{i'sj})$$
$$\forall i \in I^{in}, i' \in I^{in}, \ s \in S_i, \ j \in (J_s \cap J_i \cap J_{i'}) : i' > i \tag{8.4}$$

$$C_{i's} + \gamma_{i'ij} \leq C_{is} - \pi_{ij} - \tau_{isj} + MX_{i'ij} + M(2 - Y_{isj} - Y_{i'sj})$$
$$\forall i \in I^{in}, \ i' \in I^{in}, \ s \in S_i, \ j \in (J_s \cap J_i \cap J_{i'}) : i' > i \tag{8.5}$$

$$Y_{isj} \in \{0,1\} \quad \forall i \in I^{in}, \ s \in S_i, \ j \in (J_s \cap J_i)$$
$$X_{ii'j} \in \{0,1\} \quad \forall i \in I^{in}, \ i' \in I^{in}, \ j \in (J_s \cap J_i') : i' \neq i$$
$$W_{is} \geq 0 \quad \forall i \in I^{in}, \ s \in S_i : s < S \tag{8.6}$$
$$C_{is} \geq 0 \quad \forall i \in I^{in}, \ s \in S_i$$

Constraint set (8.1) ensures that every product order goes through one unit $j \in (J_s \cap J_i)$ at each stage $s \in S_i$. Constraint set (8.2) defines the completion time of the first stage for every product. Notice that this set of constraints takes into account possible release order $o_i$ and available unit $\varepsilon_j$ times. Constraint set (8.3) provides the timing for every product order between consecutive stages. This set of constraints allows for the consideration of possible transferring times between two sequential stages. The positive variable $W_{is-1}$ reflects the wait time of each batch product before proceeding to the following processing stage. Note that in a Zero Wait (ZW) storage policy $W_{is-1}$ is set to zero. In Unlimited Intermediate Storage (UIS) policy, $W_{is-1}$ is left free or, alternatively, it can be eliminated and the equality can be substituted by a greater-or-equal inequality. In order to model storage policies like Non-Intermediate Storage (NIS) and Finite Intermediate Storage (FIS), appropriate sets of constrains found in the literature can be easily added to the current model. Constraint sets (8.4) and (8.5) define the relative sequencing of product batches at each processing unit. These sets of big-M constraints force the starting time of a product $i'$ to be greater than the completion time of whichever product $i$ processed beforehand. Note that $X_{ii'j}$ corresponds to the global sequencing binary variable. Have in mind that $X_{ii'j}$ is active (i.e., $X_{ii'j} = 1$) for all product batches $i'$ that are processed after product batch $i$. Finally, the decision variables are defined by (8.6). Henceforth, we will refer to the MIP model that constitutes by constraint sets (8.1)–(8.6) as GP.

### 8.3.2  A Unit-Specific General Precedence Multistage Scheduling Framework

The following sets of constraints are proposed for scheduling problems where the changeover issues should be optimally integrated into the optimization framework. The constraints are as follows:

$$\sum_{j \in (J_i \cap J_s)} Y_{isj} = 1 \quad \forall i \in I^{in}, \, s \in S_i \tag{8.7}$$

$$C_{is} \geq \sum_{j \in (J_i \cap J_s)} (\max[\varepsilon_j, o_i] + \pi_{ij} + \tau_{isj}) Y_{isj}$$
$$+ \sum_{i' \neq i} \sum_{j \in (J_s \cap J_i \cap J_{i'})} \gamma_{i'ij} \overline{X}_{i'ij} \quad \forall i \in I^{in}, \, s \in S_i : s = 1 \tag{8.8}$$

$$C_{is} - \sum_{j \in (J_i \cap J_s)} (\pi_{ij} + \tau_{isj}) Y_{isj} = C_{is-1} + W_{is-1} + \mu_{s-1s}$$
$$\forall i \in I^{in}, \, s \in S_i : s > 1 \tag{8.9}$$

$$C_{is} + \gamma_{i'ij} \overline{X}_{iij} \leq C_{i's} - \pi_{i'j} - \tau_{i'sj} + M(1 - X_{ii'j})$$
$$\forall i \in I^{in}, i' \in I^{in}, s \in S_i, j \in (J_s \cap J_i \cap J_{i'}) : i' \neq i \tag{8.10}$$

$$Y_{isj} + Y_{i'sj} \leq 1 + X_{ii'j} + X_{i'ij}$$
$$\forall i \in I^{in}, \, i' \in I^{in}, \, s \in S_i, \, j \in (J_s \cap J_i \cap J_{i'}) : i' > i \tag{8.11}$$

$$2(X_{ii'j} + X_{i'ij}) \leq Y_{isj} + Y_{i'sj}$$
$$\forall i \in I^{in}, \, i' \in I^{in}, \, s \in S_i, \, j(J_s \cap J_i \cap J_{i'}) : i' > i \tag{8.12}$$

$$Z_{ii'j} = \sum_{i'' \in I^{in}: i'' \neq [i,i']} (X_{ii''j} - X_{i'i''j}) + M(1 - X_{ii'j})$$
$$\forall i \in I^{in}, \, i' \in I^{in}, \, j \in (J_i \cap J_{i'}) : i' \neq i \tag{8.13}$$

$$Z_{ii'j} + \overline{X}_{ii'j} \geq 1 \quad \forall i \in I^{in}, \, i' \in I^{in}, \, j \in (J_i \cap J_{i'}) : i' \neq i \tag{8.14}$$

$$Y_{isj} \in \{0,1\} \quad \forall i \in I^{in}, s \in S_i, j \in (J_s \cap J_i)$$
$$X_{ii'j} \in \{0,1\} \, \& \, \overline{X}_{ii'j} \in \{0,1\} \quad \forall i \in I^{in}, \, i' \in I^{in}, j \in (J_i \cap J_i^{i'}) : i' \neq i$$
$$Z_{ii'j} \in \Re \quad \forall i \in I^{in}, \, i' \in I^{in}, j \in (J_i \cap J_i^{i'}) : i' \neq i \tag{8.15}$$
$$W_{is \geq 0} \quad \forall i \in I^{in}, s \in S_i : s < S$$
$$C_{is} \geq 0 \quad \forall i \in I^{in}, s \in S_i$$

Constraint set (8.7) forces that every product order goes through one unit $j \in (J_i \cap J_s)$ at each stage $s \in S_i$. Constraint set (8.8) determines the completion time of the first stage for every product. Notice that $\overline{X}_{ii'j}$ is the unit-specific immediate precedence binary variable. Constraint set (8.9) defines the timing for every product order between to consecutive stages and is similar to constraint set (8.3) of the GP model. Constraint sets (8.10)–(8.12) define the relative sequencing of product batches at each processing unit. Big-M constraint set (8.10) forces the starting time

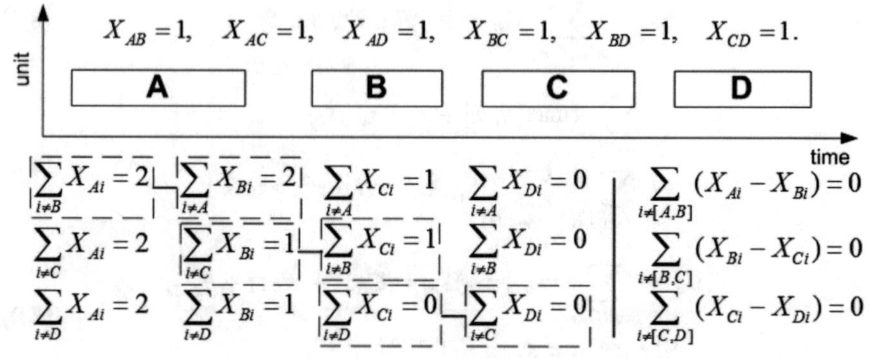

$$X_{AB} = 1, \quad X_{AC} = 1, \quad X_{AD} = 1, \quad X_{BC} = 1, \quad X_{BD} = 1, \quad X_{CD} = 1.$$

$$\sum_{i \neq B} X_{Ai} = 2 \quad \sum_{i \neq A} X_{Bi} = 2 \quad \sum_{i \neq A} X_{Ci} = 1 \quad \sum_{i \neq A} X_{Di} = 0 \quad \sum_{i \neq [A,B]} (X_{Ai} - X_{Bi}) = 0$$

$$\sum_{i \neq C} X_{Ai} = 2 \quad \sum_{i \neq C} X_{Bi} = 1 \quad \sum_{i \neq B} X_{Ci} = 1 \quad \sum_{i \neq B} X_{Di} = 0 \quad \sum_{i \neq [B,C]} (X_{Bi} - X_{Ci}) = 0$$

$$\sum_{i \neq D} X_{Ai} = 2 \quad \sum_{i \neq D} X_{Bi} = 1 \quad \sum_{i \neq D} X_{Ci} = 0 \quad \sum_{i \neq C} X_{Di} = 0 \quad \sum_{i \neq [C,D]} (X_{Ci} - X_{Di}) = 0$$

**Fig. 8.1** The unit-specific general precedence concept

of a product batch $i'$ to be greater than the completion time of whichever product batch $i$ processed beforehand at the same unit. Constraint sets (8.11) and (8.12) state that when two product batches are allocated to the same unit (i.e., $Y_{isj} = Y_{i'sj} = 1$), one of the two global sequencing binary variables $X_{ii'j}$ and $X_{i'ij}$ should be active. If the two product batches are not allocated to the same unit, then $X_{ii'j} = X_{i'ij} = 0$. It is clear that two orders $i$ and $i'$ are consecutive only in the case that $X_{ii'j} = 1$ and, moreover, when there is no other order $i''$ between them. In other words, two product batches $i$ and $i'$ are consecutive *if and only if* the total number of batches that are processed after batch $i$, if batch $i'$ is excluded, is equal to the total number of batches that are processed after batch $i'$, when batch $i$ is excluded; see Fig. 8.1. Constraint sets (8.13) and (8.14) formulate this concept. Note that the auxiliary variable $Z_{ii'j}$ is zero whenever two products $i$ and $i'$ are sequentially processed in the same unit. The RHS of constraints (8.14) can be substituted by $X_{i'ij}$; in some instances this reduces the computational time. For a more detailed description of the unit-specific general precedence concept refer to Appendix B. Finally, the decision variables are defined by (8.15). Henceforth, we will refer to the MIP model that constitutes by constraint sets (8.7)–(8.15) as USGP.

### 8.3.3   Objective Functions

In this subsection, different optimization goals for solving the short-term scheduling problem under consideration are reviewed.

**Makespan** The time point at which all product orders are accomplished corresponds to the makespan, which is calculated by Eq. (8.16). The makespan objective is closely related to the throughput objective. For instance, minimizing the makespan in a parallel-machine environment with changeover times forces the scheduler to balance the load over the various machines and to minimize the sum of all the setup times in the critical bottleneck path (Pinedo and Chao 1999).

$$\min C_{\max} \geq C_{is} \quad \forall i \in I^{\text{in}}, \ s \in S_i^{\text{last}} \tag{8.16}$$

**Total Weighted Lateness** The minimization of a combined function of earliness and tardiness, as given in Eq. (8.17), is one of the most widely used objective functions in the scheduling literature. It is also known as weighted lateness. The weighing coefficients $\alpha_i$ and $\beta_i$ are used to specify the significance of every product order earliness or tardiness, respectively.

$$\min \sum (\alpha_i E_i + \beta_i T_i) \tag{8.17}$$

Earliness and tardiness for every product order $i$ are estimated by constraint set (8.18) and (8.19), respectively.

$$E_i \geq \delta - C_{is} \quad \forall i \in I^{\text{in}}, \ s \in S_i^{\text{last}} \tag{8.18}$$

$$T_i \geq C_{is} - \delta_i \quad \forall i \in I^{\text{in}}, \ s \in S_i^{\text{last}} \tag{8.19}$$

This objective, in a sense, accounts for minimizing storage and handling costs while maximizing service and customer satisfaction level.

**Operating and Changeovers Costs** The minimization of operating and changeovers costs constitutes a reasonable goal in production environments where changeover costs are significant. The operating cost is denoted by $\psi$ and is defined as the cost for operating the production facility per time unit. Obviously, makespan corresponds to the total operating time. Parameter $\xi_{ii'j}$ stands for the changeover cost from product order $i$ to $i'$ in processing unit $j$.

$$\min \left( \psi C_{\max} + \sum_{i \in I^{\text{in}}} \sum_{i' \in I^{\text{in}}, i' \neq i} \sum_{j \in (J_i \cap J_s)} \xi_{ii'j} \overline{X}_{ii'j} \right) \tag{8.20}$$

## 8.4  The MIP-Based Solution Strategy

Although the above mathematical formulations are able to describe a large number of scheduling problems, in practice, they can only solve problems of modest size. Given that the combinatorial complexity strongly increases with the number of product orders considered, the solution of real-life industrial scheduling problems by exact methods is impossible. According to Herrmann (2006), algorithms that can find optimal solutions to these hard problems in a reasonable amount of time are unlikely to exist.

In a nutshell, the proposed MIP-based solution strategy has as a core a MIP scheduling framework and consists of two major procedure steps: (i) the constructive step, and (ii) the improvement step. The objective in the constructive step is the generation of a feasible schedule in a short amount of time. Afterwards, this

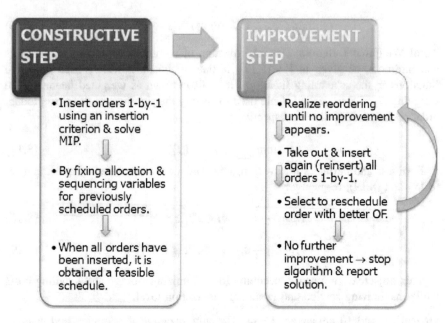

**Fig. 8.2**  Representative scheme of the proposed MIP-based solution strategy

schedule is gradually improved by implementing some elaborate rescheduling techniques, in the improvement step. As a sequence, the generation of feasible and fairly good schedules in reasonable computational times is favored. A description of the proposed solution strategy steps follows (see Fig. 8.2).

### 8.4.1   Constructive Step

In the constructive step, the large-scale scheduling problem is decomposed, in an iterative mode, into a subset of the involved product orders. This way the MIP solver search space is reduced and the resolution of the problem is favored. More specifically, a predefined number of product orders $(i \in I^{in})$ are scheduled (by solving the MIP model) at each iteration, until all product orders are finally scheduled. The user defines the number of product order per iteration. It should be noted that the number of orders inserted per iteration should be small enough to ensure a quick MIP model resolution for each iteration, and thus generating a feasible schedule in short time. In this study, it is proposed to insert (schedule) product orders one-by-one, since it has been observed, after a series of experiments, that insertion of a higher number of products per iteration: (i) does not guarantee a better constructive step solution, and (ii) is more computationally expensive.

The user should also specify the order that products are inserted into the constructive step procedure. An insertion criterion could be adopted in order to decrease the possibility of obtaining a bad constructive step solution. Here, it is proposed to insert as a priority products with less unit-stage allocation flexibility. In other words, products with less alternative units should be scheduled first. By doing so, unit allocation decisions are first taken for the less unit-stage-flexible products.

Consider a single-stage two product (A and B) batch plant with two parallel processing units (J1 and J2). The processing time of product A in unit J1 is 3 h and in unit J2 is 2 h. Product B can be only processed in unit J2 in 3 h. The minimization of makespan is the optimization goal. Consider the following insertion sequences: case I according to which product A is first inserted, which opposes our proposed insertion criterion, and case II that product B is first, which is in accordance with our proposed insertion criterion. As one can observe, the first insertion strategy (case I) results in a makespan of 5 h. Note that both products are allocated to unit J2. Following our insertion criterion (case II) a makespan of 3 h is obtained. Figure 8.3 illustrates the schedules for both cases.

After the resolution of the MIP model at each iteration, allocation and global sequencing binary variables for the already scheduled product orders are fixed. In other words, unit allocation decisions and relative sequencing relations between the already scheduled products cannot be modified in the following iterations. However, timing decisions may change thus permitting the insertion of new product orders among the previously scheduled product orders. Figure 8.4 delineates an illustrative example (single-stage products and single-unit) of the allowed sequences when a product D is inserted to a current schedule containing products A, B, and C. Note that just 4 sequences are permitted, instead of the 24 possible sequences, thus reducing significantly the computational effort. When all product orders have been inserted, a feasible schedule can be finally obtained in a relatively short time.

Similar insertion methods have also been implemented to other types of scheduling problems by Nawaz et al. (1983), Werner and Winkler (1995), Röslof et al. (2001) and Röslof et al. (2002). It is pointed out that the insertion order of product orders influences the quality of the solution. Therefore, a more detailed study and the development of other insertion criteria seems a promising future research direction for enhancing the proposed approach.

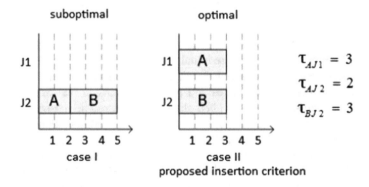

**Fig. 8.3**  Illustrative example for insertion criterion

FORBIDDEN SEQUENCES

| A | C | B | D |
|---|---|---|---|
| A | C | D | B |
| A | D | B | C |
| B | A | C | D |
| B | A | D | C |
| B | C | A | D |
| B | C | D | A |
| B | D | A | C |
| B | D | C | A |
| C | A | B | D |
| C | A | D | B |
| C | B | A | D |
| C | B | D | A |
| C | D | A | B |
| C | D | B | A |
| D | A | C | B |
| D | B | A | C |
| D | B | C | A |
| D | C | A | B |
| D | C | B | A |

ALLOWED SEQUENCES

| A | B | C | D |
|---|---|---|---|
| A | B | D | C |
| A | D | B | C |
| D | A | B | C |

☐ inserted product order

☐ previously scheduled product order

**Fig. 8.4** Illustrative example for allowed sequences in constructive step

## 8.4.2  Improvement Step

The initial feasible schedule provided by the constructive step can be systematically improved through reordering and/or reassignment MIP-based operations; in accordance with the main rescheduling concepts introduced by Röslof et al. (2001) and Méndez and Cerdá (2003a). The improvement step is a two-stage closed-loop procedure that consists of the reordering and the reinsertion stage, which are performed sequentially until no improvement is observed. A description of the improvement step follows.

**Reordering Stage** At this stage, unit allocation decisions, are fixed. Reordering actions are iteratively applied on the initial schedule, by solving a MIP model, until no further improvement is observed. A full unit reordering tactic results impractical due to the large number of batches and processing units in real-world industrial scheduling problems. Instead, the alternative of limited reordering operations may usually improve the current schedule with relatively low computational effort. It is common sense that there exists a strong trade-off between the degrees of freedom and the solution time. In an industrial environment, the scheduler should appropriately define the reordering tactic/limitations, followed in this step, depending on the complexity of the scheduling problem. A local reordering tactic is adopted in

| A | B | C | D |
|---|---|---|---|
| A | B | D | C |
| A | C | B | D |
| A | C | D | B |
| A | D | B | C |
| A | D | C | B |
| B | A | C | D |
| B | A | D | C |
| B | C | A | D |
| B | C | D | A |
| B | D | A | C |
| B | D | C | A |

| C | A | B | D |
|---|---|---|---|
| C | A | D | B |
| C | B | A | D |
| C | B | D | A |
| C | D | A | B |
| C | D | B | A |
| D | A | B | C |
| D | A | C | B |
| D | B | A | C |
| D | B | C | A |
| D | C | A | B |
| D | C | B | A |

☐ local reordering possible sequence        ■ current schedule sequence

**Fig. 8.5** Illustrative example for local reordering

this study. Thus, in an attempt to maintain manageable model sizes, reordering of batches with their direct predecessor or successor is only allowed. An illustrative example is used here to highlight the local reordering computational benefits.

Consider the reordering scheduling problem of four single-stage products (A, B, C, and D) on a single-unit. As Fig. 8.5 shows, a local reordering policy will only examine 4 potential sequences instead of the 23 total possible sequences. On the one hand, the solution quality is probably decreased since one of the 19 unexplored sequences may yield a better solution. On the other hand, the optimization search space is significantly reduced. Keep in mind that considering the whole set of possible sequences impacts drastically the computational performance of the reordering step. Other less-limited reordering tactics could be also easily applied. More details are provided in the work of Méndez and Cerdá (2003a).

**Reinsertion Stage** The schedule of the reordering step constitutes the initial schedule in the reinsertion stage. Here, unit allocation and relative sequencing decisions for a small number of product orders are left free by the scheduler. Let us refer to these product orders as reinserted orders. Allocation and relative sequencing decisions, among the non-reinserted orders, are fixed. In other words, some products orders are extracted from the current schedule, and they are reinserted aiming at improving the actual schedule. Note that the reinsertion stage is quite similar to the last iteration of the constructive step (see Fig. 8.4). Since our scope is to propose a general standard algorithm for large-scale industrial scheduling problems, we adopt the lowest number of reinsertion orders (i.e., one at a time) in order to favor low solution times. However, the scheduler could set the number of reinserted orders depending on the specific scheduling problem.

In the standard reinsertion stage, the number of iterations (reinsertions) equals the number of product orders. The solutions of all reinserted orders (iterations) are compared, and the best one is finally chosen as the solution of the reinsertion stage.

Note that if the number of product orders is too high, someone could have preferred to end the reinsertion stage once a better solution (comparing it with the previous stage) is reached. This way computational savings are achieved. If the best solution at this stage is better than the solution of the reordering stage, the algorithm goes to the reordering stage again. Otherwise, the solution algorithm terminates and reports the best solution found.

In Appendix E, some illustrative pseudocodes can be found for the constructive and the improvement stage of our MIP-based solution strategy.

## 8.5    Pharmaceutical Production Process

A real-world multiproduct multistage pharmaceutical batch plant is studied in the current work. Recently, Castro et al. (2009) have also studied this pharmaceutical facility. More specifically, they solved two problem instances (for 30 and 50 product orders) minimizing the makespan under UIS policy. In this work, we use partially different sets of data (e.g., we introduce due dates, changeover costs) and we deal with more objective functions.

In this study, the short-term scheduling problem of a considerably high number of multistage product orders (30 and 60) using 17 processing units in the production plant is addressed. The production process has six processing stages, as Fig. 8.6 depicts. Some products bypass the third processing stage S3. Changeover times are also explicitly considered thus increasing the complexity of the problem. An interesting feature of the production process is that in some processing stages changeover times are higher than the processing times. Changeover times are zero in the first stage S1 and 0.45 h in the second stage S2 among all products. Changeover times for the remaining stages (S3–S6) and processing times can be found in Appendix E. Finally, changeover costs are defined as the multiplication of the impact factors given in Table 8.1 and the corresponding changeover time.

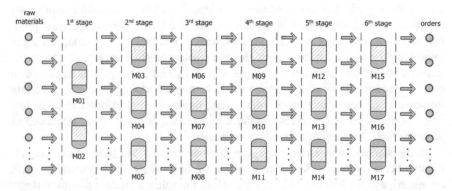

**Fig. 8.6** Pharmaceutical multistage process

**Table 8.1** Changeover costs' impact factors per time unit (103 $/h)

| Products | P01–P10 | P11–P20 | P21–P30 |
|----------|---------|---------|---------|
| P01–P10  | 0.36    | 0.27    | 0.27    |
| P11–P20  | 0.27    | 0.45    | 0.27    |
| P21–P30  | 0.27    | 0.27    | 0.54    |

## 8.6 Experimental Studies

In this section, the problem instances details are first introduced and the results of these experimental studies are presented and discussed afterward.

### 8.6.1 Details of Problem Instances

Twelve different problem instances have been solved. These case studies differ in: (i) the optimization goal (makespan, weighted lateness, and operating and changeover costs), (ii) the number of product orders (30 products, and 60 products), and (iii) the storage policy type (ZW, and UIS).

Notice that two batches for every product are considered in order to address the 60- product cases. Therefore, for instance, the product order P31 has the same processing characteristics with product order P01, where product order P32 has the same processing data with product order P02, and so on. Moreover, notice that the changeover times/costs between P01 and P31 are equal to the changeover times/costs, as they are given in the data tables, and so on. For the problem instances where the optimization goal is the minimization of weighted lateness (i.e., PI.05–PI.08), due dates for every product order are considered, according to Table 8.2. It should be noted that due dates for two batches of the same product (e.g., P01 and P31) may be different. Additionally, the weighing coefficient for earliness, $\alpha_i$, equals to 0.9 and the weighing coefficient for tardiness, $\beta_i$, is set to 4.5 for all products. Regarding the problem instances with objective the simultaneous minimization of operating and changeovers costs (i.e., PI.09–PI.12), the operating cost per time unit, is equal to $0.9 \times 10^3$ $/h.

At this point, it is worth mentioning that the GP model has been used to solve the problem instances that involves the minimization of the makespan or the weighted lateness (i.e., PI.01–PI.08), and USGP model has been employed to cope with the operating and changeovers costs objective (i.e., PI.09–PI.12). Moreover, it is emphasized that the 30-product problem instances (PI.01, PI.02, PI.05, PI.06, PI.09, and PI.10) deal with the complex scheduling of 168 product batches, and the 60-product problem instances (PI.03, PI.04, PI.07, PI.08, PI.11 and PI.12) tackle the intricate scheduling problem of 336 product batches.

**Table 8.2** Due dates for product orders (h)

| P01 | 30.6 | P016 | 20.7 | P31 | 34.2 | P46 | 54.0 |
|-----|------|------|------|-----|------|-----|------|
| P02 | 16.2 | P017 | 14.4 | P32 | 37.8 | P47 | 49.5 |
| P03 | 23.4 | P018 | 23.4 | P33 | 37.8 | P48 | 46.8 |
| P04 | 18.0 | P019 | 20.7 | P34 | 48.6 | P49 | 52.2 |
| P05 | 27.0 | P020 | 27.0 | P35 | 40.5 | P50 | 54.0 |
| P06 | 16.2 | P021 | 30.6 | P36 | 34.2 | P51 | 48.6 |
| P07 | 28.8 | P022 | 9.0  | P37 | 46.8 | P52 | 52.2 |
| P08 | 20.7 | P023 | 18.0 | P38 | 54.0 | P53 | 27.0 |
| P09 | 18.0 | P024 | 23.4 | P39 | 46.8 | P54 | 52.2 |
| P010| 0.0  | P025 | 23.4 | P40 | 49.5 | P55 | 41.4 |
| P011| 10.0 | P026 | 18.0 | P41 | 52.2 | P56 | 49.5 |
| P012| 8.0  | P027 | 14.4 | P42 | 40.5 | P57 | 54.0 |
| P013| 30.6 | P028 | 9.0  | P43 | 54.0 | P58 | 52.2 |
| P014| 14.4 | P029 | 18.0 | P44 | 36.0 | P59 | 40.5 |
| P015| 27.0 | P030 | 10.8 | P45 | 52.2 | P60 | 36.0 |

## 8.6.2  Results and Discussion

The proposed solution strategy has been tested on a total number of 12 complex problem instances in order to validate its performance. A time limit of 1 CPU h has been imposed on the solution of every problem instance. All problem instances have been solved in a Dell Inspiron 1520 2.0 GHz with 2 GB RAM using CPLEX 11 via a GAMS 22.8 interface.

Table 8.3 presents the constructive step's solution (initial solution) and the best solution found for each problem instance. The computational time for the constructive step (First-stage) as well as the total computation time is also included in the same table. Note that feasible schedules are obtained in short computational times. Problem instance PI.11 is the most time demanding, since almost half a CPU h was needed in order to obtain a feasible solution. The remaining problem instances reached a feasible solution in relatively low computational times ranging from some CPUs to no more than 7 CPU min.

The MIP-based solution strategy is able to quickly generate feasible solutions and then gradually improve the quality these solutions. It was observed that the necessary computational time to improve a given solution mainly depends on: (i) the total number of batches to be scheduled, (ii) the objective function, (iii) the storage policy, and (iv) the core mathematical model. Obviously, the lower the total number of batches the faster the problem is solved. It has also been observed that the case studies considering ZW storage policy are solved faster comparing them with the cases under UIS policy. The MIP model used depends on the optimization goal. Generally, the more complex the objective function the bigger the size of the model; such is the case of minimizing operating and changeovers costs.

**Table 8.3** Problem instances: best schedules found within the maximum predefined time limit (3600 CPUs)

| Problem instance | Objective function | Products (batches) | Storage policy | First-stage solution | First-stage CPUs | Best solution | Total CPUs | Improvement (%) |
|---|---|---|---|---|---|---|---|---|
| PI.01 | $C_{max}$ | 30 (168) | UIS | 28.507 | 38 | 542 | 542 | 6.83 |
| PI.02 | $C_{max}$ | 30 (168) | ZW | 31.250 | 7 | 187 | 187 | 3.14 |
| PI.03 | $C_{max}$ | 60 (336) | UIS | 49.161 | 155 | 1502 | 1502 | 1.25 |
| PI.04 | $C_{max}$ | 60 (366) | ZW | 58.104 | 106 | 1718 | 1718 | 3.52 |
| PI.05 | W.L. | 30 (168) | UIS | 48.613 | 22 | 720 | 720 | 60.74 |
| PI.06 | W.L. | 30 (168) | ZW | 115.016 | 15 | 262 | 262 | 26.59 |
| PI.07 | W.L. | 60 (336) | UIS | 118.683 | 403 | 3600 | 3600 | 25.90 |
| PI.08 | W.L. | 60 (336) | ZW | 629.672 | 356 | 1478 | 1478 | 18.07 |
| PI.09 | O.C.C. | 30 (168) | UIS | 66.158 | 94 | 3600 | 3600 | 4.91 |
| PI.10 | O.C.C. | 30 (168) | ZW | 75.318 | 58 | 3600 | 3600 | 2.92 |
| PI.11 | O.C.C. | 60 (336) | UIS | 119.759 | 1780 | 3600 | 3600 | 1.54 |
| PI.12 | O.C.C. | 60 (336) | ZW | 139.104 | 880 | 3600 | 3600 | 3.22 |

W.L.—Weighted lateness, and O.C.C.—operating and changeovers costs

All problem instances were solved using the original undecomposed mathematical formulations in order to underline the high complexity of the problems addressed and to highlight the practical benefits of the proposed solution approach. Table 8.4 contains the computational features of the original mathematical models and the best solution found, for all problem instances, within the predefined time limit of 1 CPU h. It is worth mentioning that the number of the sequencing binary variables is strongly augmented by increasing the number of product orders. Note that the least complex problem instances (PI.01 and PI.02) result into a MIP model of 10,230 equations, 5326 binary variables, and 295 continuous variables, while, the most complex problem instances (PI.11 and PI.12) result into a huge MIP model of 161,828 constraints, 41,056 binary variables, and 81,626 continuous variables. It should be noted that the original mathematical models in 8 of the 12 problem instances did not find a feasible solution. In the remaining problem instances, feasible but very bad solutions were obtained. For example, the original GP model in problem instance PI.01 reported a makespan equal to 34.810 h, with an integrality gap of 58%, after 1 CPU h while our solution approach gave a makespan of 26.559 h in just 542 CPUs. The solution found by the original GP model is 31.07% worse than that of our approach. The original MIP models also solved all problem instances without setting a time limit. However, in all cases, the MIP solver terminated because memory capacity was exceeded.

According to Table 8.4, it is evident that the proposed MIP-based solution strategy overwhelms the original MIP models. Using our approach, highly complicated scheduling problems in multiproduct multistage batch plants can be solved, as the current experimental study reveals. Although optimality cannot be guaranteed in general, feasible solutions can be obtained in relatively short computational

**Table 8.4** Comparison between the original MIP model and the proposed MIP-based strategy best solutions found within the maximum predefined time limit (3600 CPUs)

| Original mathematical model | | | | | | | Our strategy | |
|---|---|---|---|---|---|---|---|---|
| Problem instance | Constraints | Binary variables | Continuous variables | Gap (%) | Best solution | Total CPUs | Best solution | Total CPUs |
| PL.01 | 10,230 | 5326 | 295 | 58 | 34.810 | 3600 | 26.559 | 542 |
| PL.02 | 10,230 | 5326 | 295 | – | – | 3600 | 30.552 | 187 |
| PL.03 | 40,998 | 20,916 | 589 | 90 | 109.960 | 3600 | 48.548 | 1502 |
| PL.04 | 40,998 | 20,916 | 589 | – | – | 3600 | 56.061 | 1718 |
| PL.05 | 10,261 | 5326 | 355 | 100 | 428.146 | 3600 | 19.085 | 720 |
| PL.06 | 10,261 | 5326 | 355 | – | – | 3600 | 84.438 | 262 |
| PL.07 | 41,049 | 20,916 | 709 | 100 | 23,453.744 | 3600 | 87.943 | 3600 |
| PL.08 | 41,049 | 20,916 | 709 | – | – | 3600 | 515.876 | 1478 |
| PL.09 | 39,858 | 10,264 | 20,286 | – | – | 3600 | 62.910 | 3600 |
| PL.10 | 39,858 | 10,264 | 20,286 | – | – | 3600 | 70.209 | 3600 |
| PL.11 | 161,828 | 41,056 | 81,626 | – | – | 3600 | 117.909 | 3600 |
| PL.12 | 161,828 | 41,056 | 81,626 | – | – | 3600 | 134.624 | 3600 |

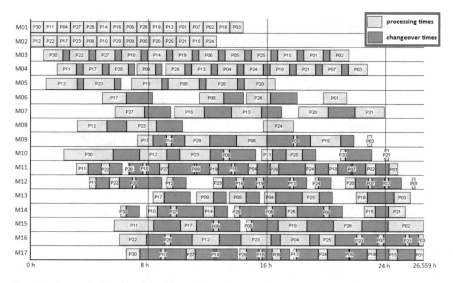

**Fig. 8.7** Best schedule for PI.01 (30-product case: minimization of makespan under UIS policy)

time. Bear in mind that feasibility is the principal goal in practical scheduling problems. To the best of our knowledge, neither other standard solution methods nor heuristics exist for tackling the studied scheduling problem efficiently.

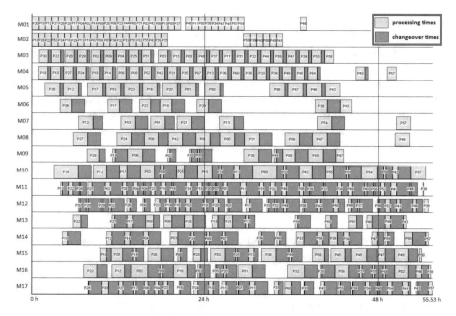

**Fig. 8.8** Best schedule for PI.07 (60-product case: minimization of total weighted lateness under UIS policy)

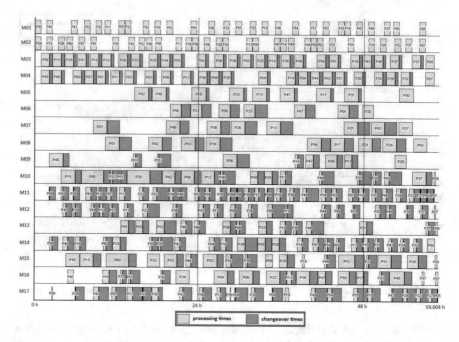

**Fig. 8.9** Best schedule for PI.12 (60-product case: minimization of total operating and changeovers costs under ZW policy)

Some Gantt charts of the best schedules for some representative problem instances are provided in order to provide the reader with a visual demonstration of the complexity of the addressed problems. More specifically, Fig. 8.7 presents the best solution found for solving the 30-product case by minimizing $C_{max}$ under UIS policy (problem instance PI.01). Figure 8.8 illustrates the best schedule reported for solving the 60-product case by minimizing total weighted lateness under UIS policy (problem instance PI.07). Finally, Fig. 8.9 graphically depicts the best schedule found for solving the 60-product case by minimizing total operating and change-overs costs under ZW storage policy (problem instance PI.12).

## 8.7  Concluding Remarks

A novel iterative two-step MIP-based solution strategy has been presented for the solution of large-scale scheduling problems in multiproduct multistage batch plants. A benchmark scheduling problem in a multiproduct multistage pharmaceutical batch plant has been introduced and solved in this study. The proposed solution technique is able to generate good feasible solutions in relatively short times, as the several problem instances of the pharmaceutical scheduling problem reveal. It is worthwhile to note that the user can appropriately define the degrees of freedom of the decision variables by balancing the trade-off between computational time and solution quality. The proposed solution strategy can be also applied to other types of scheduling problems by adopting a different MIP core model that describes the particular scheduling problem. Moreover, this work aims to be a step toward reducing the gap between scheduling theory and practice, since it has clearly demonstrated that effective MIP-based optimization solution strategies can solve real-world industrial scheduling problems (Kopanos et al. 2009).

## 8.8 Nomenclature

**Indices/Sets**

$i, i', i'' \in I$    Product orders (products)
$j \in J$         Processing units (units)
$s \in S$        Processing stages (stages)

**Subsets**

$I^{\text{in}}$   Set of products $i$ that are included into the optimization
$J_i$   Available units $j$ to process product $i$
$J_s$   Available units $j$ to process product $i$
$S_i$   Set of stages $s$ for each product order $i$
$S_i^{\text{last}}$   Last processing stage for product order $i$

**Parameters**

$\alpha_i$   Weighing coefficient for earliness for product $i$
$\beta_i$   Weighing coefficient for tardiness for product $i$
$\gamma_{ii'j}$   Sequence-dependent setup (changeover) time between products $i$ and $i'$ in unit $j$
$\delta_i$   Due date for product $i$
$\varepsilon_j$   Time point that unit $j$ is available to start processing
$M$   A big number
$\mu_{s-1s}$   Batch transfer time between two consecutive stages $s-1$ and $s$
$\xi_{ii'j}$   Sequence-dependent setup (changeover) cost between products $i$ and $i'$ in unit $j$
$o_i$   Release time for product $i$
$\pi_{ij}$   Sequence-independent setup time of product $i$ in unit $j$
$\tau_{isj}$   Processing time for stage $s$ of product $i$ in unit
$\psi$   Operating cost of production facility per time unit

**Continuous Variables**

$C_{is}$   Completion time of stage $s$ of product $i$
$C_{\text{max}}$   Makespan
$E_i$   Earliness for product $i$

$T_i$      Tardiness for product $i$

$W_{is}$   The time that stage $s$ of a product $i$ is stored (waits) before proceeding to the following processing stage $s +1$

$Z_{ii'j}$  Allocation position difference between products $i$ and $i'$ in unit $j$

**Binary Variables**

$X_{ii'j}$  = 1, for every product $i$ that is processed before product $i'$ in unit $j$

$\overline{X}_{ii'j}$  = 1, if product $i$ is processed exactly before product $i'$ in unit $j$

$Y_{isj}$  = 1, if stage $s$ of product $i$ is assigned to unit $j$

# References

Castro PM, Harjunkoski I, Grossmann IE (2009) Optimal short-term scheduling of large-scale multistage batch plants. Ind Eng Chem Res 48:11002–11016

Hermann JW (2006) Handbook of production scheduling. Springer Science+Business Media, Inc., New York

Kopanos GM, Laínez JM, Puigjaner L (2009) An efficient mixed-integer linear programming scheduling framework for addressing sequence-dependent setup issues in batch plants. Ind Eng Chem Res 48(13):6346–6357

Kopanos GM, Méndez CA, Puigjaner L (2010) MIP-based decomposition strategies for large-scale scheduling problems in multiproduct multistage batch plants: a benchmark scheduling problem of the pharmaceutical industry. Eur J Oper Res 207(2):644–655

Méndez CA, Cerdá J (2003) Dynamic scheduling in multiproduct batch plants. Comput Chem Eng 27:1247–1259

Méndez CA, Cerdá J et al (2006) Review: state-of-theart of optimization methods for short-term scheduling of batch processes. Comput Chem Eng 30(6–7):913–946

Nawaz M, Enscore EE, Ham I (1983) A heuristic algorithm for the m-machine n-job flowshop sequencing problem. OMEGA, Int J Manag Sci 11:91–95

Pinedo M, Chao X (1999) Operations scheduling with applications in manufacturing and services. McGraw-Hill International Editions, New York

Röslof J, Harjunkoski I et al (2001) An MILP-based reordering algorithm for complex industrial scheduling and rescheduling. Comput Chem Eng 25:821–828

Röslof J, Harjunkoski I et al (2002) Solving a large-scale industrial scheduling problem using MILP combined with a heuristic procedure. Eur J Oper Res 138:29–42

Ruiz R, Serifoglu FS, Urlings T (2008) Modeling realistic hybrid flexible flowshop scheduling problems. Comput Oper Res 35:1151–1175

Werner F, Winkler A (1995) Insertion techniques for the heuristic solution of the job shop problem. Discrete Appl Math 58:191–211

# Part V
# Integrated-System Approach

# Chapter 9
# Integrated Operational and Maintenance Planning of Production and Utility Systems

## 9.1 Introduction

Previous chapters have been dedicated to a new conception of planning and detailed production scheduling in the process industries, thus improving their key operational activities (Kopanos et al. 2009, 2010a, b). This chapter focuses on one of the main goals of any process industry, which is to generate maximum revenues at low costs by maintaining high production levels in order to satisfy the demand for products. A means for achieving this is by following a plant-wide approach through the integrated management of operational and maintenance tasks in the overall process system (Zulkafli and Kopanos 2016, 2017).

Major industrial facilities consist of interconnected production and utility systems.

Figure 9.1 displays a representative layout of production and utility systems for a process industry. Under this plant layout, the production system produces desired products from raw materials that may undergo several production processes, such as reactions or separations. These main production processes require large amounts of different utilities, such as power, steam, compressed air, industrial gases, or water. Especially, energy-intensive process industries have an onsite utility system that generates the major utilities required by the main production system. Combined heat and power units, gas or steam turbines, compressors, and boilers are examples of onsite utility systems. The raw materials of the utility system can be any type of fuel or other resource such as atmospheric air or water. These materials undergo a conversion process in utility units to generate the desired utilities. Depending on the type of utility, chemical or physical conversion could take place in a utility unit (e.g., combustion or compression). Then, the generated utilities are supplied to the production system for its own operation and the production of intermediate or final products. Excessive amounts of utilities can be stored in buffer tanks (e.g., hot water), be recycled (e.g., steam), or in some cases be released to the environment

© Springer Nature Switzerland AG 2019
G. M. Kopanos and L. Puigjaner, *Solving Large-Scale Production Scheduling and Planning in the Process Industries*,
https://doi.org/10.1007/978-3-030-01183-3_9

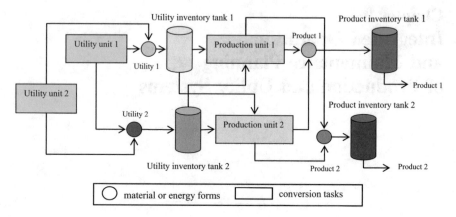

**Fig. 9.1** Representative layout for the interaction of production and utility systems

(e.g., exhaust heat). Some utilities may be acquired from external sources under an associated cost, if the onsite utility system cannot meet the needs of the production system (e.g., electricity from the power grid). utility units may. Final products or utilities can be stored in dedicated inventory tanks or directly satisfy the demand for products or the utility requirements of the production system, respectively.

In addition to the above, modern process plants consist of complex operating equipment that require maintenance to perform its required function in a timely manner to avoid equipment damage and inefficient use. Effective maintenance policies can sustain the operational level, reduce operating costs, and restrain the equipment and the overall system from entering hazardous states. The cleaning of production or utility equipment that are subject to performance degradation is one of the major maintenance actions in process industries. The purpose of this cleaning is to recover the performance (efficiency) of the corresponding equipment and decrease energy consumption over its operation. Thus, it is essential to consider condition-based maintenance policies for the equipment of a process plant to increase its overall energy efficiency, operability, and stability (Xenos et al. 2016; Kopanos et al. 2015). To do this, performance degradation and recovery models need to be derived for each equipment and alternative maintenance policies need to be considered (e.g., online or offline cleaning).

Nowadays, process industries typically follow a sequential approach for the optimization of the operational plan of their production and utility systems. The main drawback of this approach is that it provides suboptimal solutions (with respect to energy efficiency and costs) since the two interconnected systems are not optimized simultaneously. Importantly, this traditional approach often faces the risk of providing generation targets for utilities that cannot be met by the utility system (infeasible solutions), and in that case either purchase of utilities would take place or a re-planning of the production may be needed (Zulkafli and Kopanos 2016;

Kopanos and Pistikopoulos 2014). Additionally, maintenance of production or utility units are typically predefined or follow a very conservative plan and not optimized by considering the actual operational plan of the overall process system.

Comparisons with solutions obtained by using a sequential approach indicated that the integrated approach leads to a significant reduction in energy costs and at the same time decreases the emissions of harmful gases. In essence, a systematic approach is needed for addressing the plant-wide management and planning of a process industry .... This integrated approach is a key step toward the transformation of current process industries to smart process industries, following the Internet-of-Things revolution, where all operations are performed to achieve substantially enhanced energy, sustainability, and environmental and economic performance.

Further, in this chapter, comprehensive comparisons are made between the solutions obtained following the proposed integrated approach and the traditional sequential approach, demonstrating clearly the benefits of the proposed approach over its sequential counterpart. Overall, the proposed integrated method follows a whole-system approach that addresses the efficient energy generation, use and consumption (i.e., production and utility units under performance degradation and recovery), improved material handling (i.e., resource-constrained cleaning policies), and integrated management of energy and material resources in dynamic environments (i.e., integrated approach under uncertainties) towards a cleaner and sustainable production in process industries.

## 9.2 Problem Statement

This work focuses on the detailed condition-based operational and cleaning planning of production and utility systems under alternative resource-constrained cleaning policies, through the consideration of performance degradation and recovery for utility and production units. This integrated planning problem is formally defined in terms of the following items:

- A given planning horizon divided into a number of equally length time periods $t \in T$.
- A set of energy or material resources $e \in E$ that are classified to final product $(e \in E^{PR})$ and utility resources $(e \in E^{UT})$. The final products have known demand profiles $\zeta_{(e,t)}$.
- A set of units $i \in I$ that could produce a number of resources $e \in E_i$. These units are categorized to utility $(i \in UT_i)$ and production $(i \in PR_i)$ units. Maximum (minimum) operating levels $\kappa_{(i,t)}^{max} \left( \kappa_{(i,t)}^{min} \right)$ for utility units and production levels $\bar{\kappa}_{(i,e,t)}^{max} \left( \bar{\kappa}_{(i,e,t)}^{min} \right)$ for production units are known. For the units that have a

maximum runtime ($i \in \mathrm{MR}_i$), the maximum runtime ($o_i$) after its last startup is defined. For every unit that is subject to startup and shutdown actions ($i \in I^{\mathrm{SF}}$), the startup $\left(\phi^{\mathrm{S}}_{(i,t)}\right)$ and shutdown $\left(\phi^{\mathrm{F}}_{(i,t)}\right)$ costs are also given. For any unit that is subject to minimum runtime and shutdown time restrictions (i.e., $i \in I^{\mathrm{S-min}}$ and $i \in I^{\mathrm{F-min}}$, respectively), the minimum runtime after its last startup $\omega_i$ and the minimum idle time after its last shutdown $\psi_i$ are also defined.

- A set of resource-dedicated inventory tanks $z \in Z_e$ that can receive resources from units $i \in I_z^+$ and send resources to units $i \in I_z^-$. The inventory tanks have a given maximum (minimum): inventory tank level $\beta^{\mathrm{max}}_{(e,z)}\left(\beta^{\mathrm{min}}_{(e,z)}\right)$, inlet resource flow $\beta^{+,\mathrm{max}}_{(e,z,t)}\left(\beta^{+,\mathrm{min}}_{(e,z,t)}\right)$, and outlet utility resource flow $\beta^{-,\mathrm{max}}_{(e,z,t)}\left(\beta^{-,\mathrm{min}}_{(e,z,t)}\right)$. Initial inventory tank levels $\tilde{\beta}_{(e,z)}$ and losses coefficients $\beta^{\mathrm{loss}}_z$ are also given.

- Different cleaning policies for the units are considered. In particular, a unit could be subject to: (i) flexible time-window offline cleaning ($i \in \mathrm{FM}_i$) with a given earliest $\tau^{\mathrm{es}}_i$ and latest $\tau^{\mathrm{ls}}_i$ starting time, (ii) in-progress offline cleaning carried over from the previous planning horizon ($i \in \mathrm{DM}_i$), or (iii) condition-based cleaning ($i \in \mathrm{CB}_i$) with known performance degradation rates. Two types of condition-based cleaning tasks are considered, namely: online cleaning tasks $\left(\mathrm{CB}^{\mathrm{on}}_i\right)$ with given recovery factors $\rho^{\mathrm{rec}}_i$, and offline cleaning tasks $\left(\mathrm{CB}^{\mathrm{off}}_i\right)$.

- A set of alternative cleaning tasks options $q \in Q_i$ for each unit that is subject to flexible time-window cleaning ($i \in \mathrm{FM}_i$) or offline condition-based cleaning $\left(i \in \mathrm{CB}^{\mathrm{off}}_i\right)$. The cleaning tasks options are characterized by different durations $\nu_{(i,q)}$, cleaning resource requirements $\vartheta^{\mathrm{off}}_{(i,q)}$, and associated cleaning costs $\phi^{\mathrm{off}}_{(i,q,t)}$.

- For every production unit $i \in I^{\mathrm{PR}}$, fixed and variable utility requirements for the production of final products are given ($\bar{\alpha}_{(i,e,e')}$ and $\alpha_{(i,e,e')}$, respectively).

- Given variable and fixed operating costs for production and utility units, $\phi^{\mathrm{PR,op-var}}_{(i,e,t)}$ and $\phi^{\mathrm{PR,op-fix}}_{(i,e,t)}$, and $\phi^{\mathrm{UT,op-var}}_{(i,t)}$ and $\phi^{\mathrm{UT,op-fix}}_{(i,t)}$, respectively.

- Given purchase prices for acquiring utility and product resources from external sources, $\phi^{\mathrm{UT,ex}}_{(e,i,t)}$ and $\phi^{\mathrm{PR,ex}}_{(e,t)}$ respectively.

- A given time-varying energy price profile $\phi^{\mathrm{pw}}_{(i,t)}$.

Some additional considerations of the problem under study are the following: (i) the demands for final products should be fully satisfied; and (ii) there is a limited amount of available resources for cleaning tasks per time period.

For every time period, the key decisions to be made by the optimization model are

- the operational status for each production and utility unit (i.e., startup, shutdown, in operation, idle, under cleaning);

- the operating level for each production and utility unit;
- the inventory level for each inventory tank of utility and product resources;
- the utility requirements of each production unit; and
- the selection of the timing and the types of the cleaning tasks to be performed in each production and utility unit.

And all these with the goal to minimize the cost of the overall process system which includes the following:

- fixed and variable operating costs for production and utility units;
- startup and shutdown costs for production and utility units;
- extra energy costs due to performance degradation for production and utility units;
- cleaning costs for production and utility units; and
- Penalties or costs for acquiring utility and product resources from external sources.

## 9.3 Optimization Framework

In this section, a linear MIP model is presented for the integrated planning problem considered in this study (Kopanos et al. 2012a, b). The proposed mathematical model follows a rolling horizon modeling representation in order to readily deal with various types of uncertainty, such as fluctuations on the demand for final products, unit breakdowns, variations of cost terms, or data inaccuracies. In brief, in the rolling horizon scheme, a planning problem is solved for a certain length of time horizon (i.e., prediction horizon), and then the solution for a part of that time horizon (i.e., control horizon) is executed (typically for the first time period of the prediction horizon). After each iteration, a new planning problem is solved by moving forward the time horizon by the length of the control horizon considered. Figure 9.2 displays a representative rolling horizon example for the reactive planning problem described above. In a rolling horizon framework, the state of the overall system and the uncertain parameters of the problem are updated before each iteration. The main parameters that need to be updated are: (i) the level of every inventory tank; (ii) the cumulative time of operation per unit; (iii) the deviation of the operating level per unit; (iv) the current operating status of each unit; (v) the startup and shutdown history of units; (vi) the online cleaning history of units; and (vii) the demands for products. Figure 9.3 shows a schematic representative of the steps of the proposed reactive planning method. A description of the proposed optimization framework follows (Kopanos et al. 2011a, b, c).

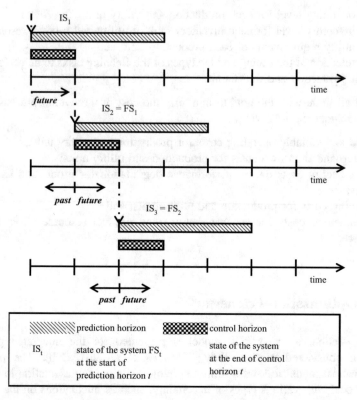

**Fig. 9.2** A representative rolling horizon example for reactive planning

## 9.3.1 Startup and Shutdown Actions

In order to model the major operational status (i.e., in operation, idle, startup, or shutdown) of production and utility units, the following set of binary variables is introduced:

$$X_{(i,t)} = \begin{cases} 1 & \text{if unit } i \text{ is operating during time period } t, \\ 0 & \text{otherwise.} \end{cases}$$

$$S_{(i,t)} = \begin{cases} 1 & \text{if unit } i \text{ starts up at the beginning of time period } t, \\ 0 & \text{otherwise.} \end{cases}$$

$$F_{(i,t)} = \begin{cases} 1 & \text{if unit } i \text{ shuts down at the beginning of time period } t, \\ 0 & \text{otherwise.} \end{cases}$$

The operational status of each unit is then modeled according to

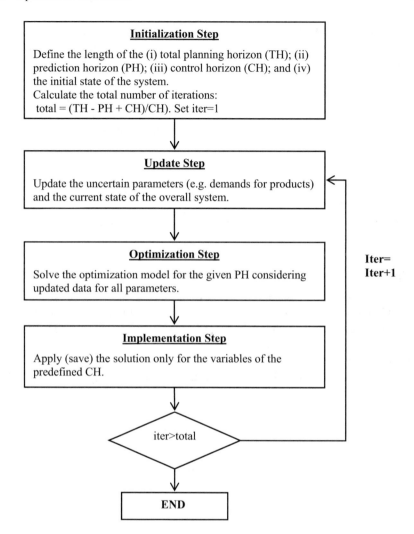

**Fig. 9.3** Reactive planning method via rolling horizon

$$
\begin{array}{ll}
S_{(i,t)} - F_{(i,t)} = X_{(i,t)} - \tilde{\chi}_i & \forall i \in I^{\mathrm{SF}}, t \in T : t = 1 \\
S_{(i,t)} - F_{(i,t)} = X_{(i,t)} - X_{(i,t-1)} & \forall i \in I^{\mathrm{SF}}, t \in T : t > 1 \qquad (9.1) \\
S_{(i,t)} + F_{(i,t)} \leq 1 & \forall i \in I^{\mathrm{SF}}, t \in T
\end{array}
$$

The first two sets of constraints relate the startup and shutdown actions with the operating binary variables, while the last set of constraints ensure that no startup and shutdown action can occur simultaneously.

The minimum runtime $\omega_i$ and shutdown time $\psi_i$ for any unit subject to minimum runtime or shutdown restriction are modeled by constraints (9.2) and (9.3), respectively.

$$
\begin{aligned}
X_{(i,t)} &\geq \sum_{t'=\max\{1,t-\omega_i+1\}}^{t} S_{(i,t')} \quad \forall i \in I^{\text{S-min}}, t \in T : \omega_i > 1 \\
X_{(i,t)} &= 1 \quad\quad\quad\quad\quad\quad\quad \forall i \in I^{\text{S-min}}, t = 1,\ldots,(\omega_i - \tilde{\omega}_i) : 0 < \tilde{\omega}_i < \omega_i
\end{aligned}
\tag{9.2}
$$

$$
\begin{aligned}
1 - X_{(i,t)} &\geq \sum_{t'=\max\{1,t-\psi_i+1\}}^{t} F_{(i,t')} \quad \forall i \in I^{\text{F-min}}, t \in T : \psi_i > 1 \\
X_{(i,t)} &= 0 \quad\quad\quad\quad\quad\quad\quad \forall i \in I^{\text{F-min}}, t = 1,\ldots,(\psi_i - \tilde{\psi}_i) : 0 < \tilde{\psi}_i < \psi_i
\end{aligned}
\tag{9.3}
$$

Parameters $\tilde{\omega}_i \left( \tilde{\psi}_i \right)$ describe the initial state of each unit with respect to its total number of consecutive operating (idle) periods since its last startup (shutdown) at the beginning of the current planning horizon. Constraints (9.2) and (9.3) are needed only if the minimum runtime $\omega_i$ or shutdown time $\psi_i$ of a unit is greater than a single time period, respectively.

Generally speaking, a maximum runtime $(o_i)$ may be imposed for units $i \in \text{MR}_i$ that do not follow a more detailed performance-based cleaning planning, according to

$$
\begin{aligned}
\sum_{t'=\max\{1,t-o_i\}}^{t} X_{(i,t')} &\leq o_i \quad\quad\quad\quad\quad \forall i \in \text{MR}_i, t \in T \\
\sum_{t'=\max\{1,t-(o_i-\tilde{\omega}_i)\}}^{t} X_{(i,t')} &\leq (o_i - \tilde{\omega}_i) \quad \forall i \in \text{MR}_i, t = (o_i - \tilde{\omega}_i + 1) : \tilde{\omega} > 1
\end{aligned}
\tag{9.4}
$$

### 9.3.2   Cleaning Tasks

As discussed in Problem Statement, the different unit cleaning policies considered are: (i) flexible time-window offline cleaning $(i \in \text{FM}_i)$, (ii) in-progress offline cleaning carried over from the previous planning horizon $(i \in \text{DM}_i)$, or (iii) condition-based cleaning $(i \in \text{CB}_i)$. Online cleaning $\left( \text{CB}_i^{\text{on}} \right)$ and offline cleaning tasks $\left( \text{CB}_i^{\text{off}} \right)$ are considered for the condition-based cleaning. The following binary variables are defined to model these cleaning tasks.

$$H_{(i,q,t)} = \begin{cases} 1 & \text{if a cleaning task option } q \text{ for } i \in (\text{CB}_i^{\text{off}} \cup \text{FM}_i) \\ & \text{begins at the start of time period } t, \\ 0 & \text{otherwise.} \end{cases}$$

$$W_{(i,t)} = \begin{cases} 1 & \text{if an offline cleaning task for } i \in (\text{CB}_i^{\text{off}} \cup \text{FM}_i) \\ & \text{begins at the start of time period } t, \\ 0 & \text{otherwise.} \end{cases}$$

$$V_{(i,t)} = \begin{cases} 1 & \text{if an online cleaning task for } i \in (\text{CB}_i^{\text{on}} \cap \text{UT}_i) \\ & \text{takes place in time period } t, \\ 0 & \text{otherwise.} \end{cases}$$

$$V_{(i,e,t)}^{\text{PR}} = \begin{cases} 1 & \text{if an online cleaning task for } i \in (\text{CB}_i^{\text{on}} \cap \text{PR}_i) \\ & \text{that produces product } e \in E_i \text{ in time period } t, \\ 0 & \text{otherwise.} \end{cases}$$

### 9.3.3   In-Progress Offline Cleaning Tasks

At the beginning of the planning horizon, there may be some in-progress unfinished offline cleaning tasks for some units $(i \in \text{DM}_i)$, which are carried over from the previous planning horizon. These cleaning tasks are modeled according to

$$X_{(i,t)} = 0 \quad \forall i \in \text{DM}_i, t \in T : \tilde{\eta}_{(i,t)} > 0 \tag{9.5}$$

Parameters $\tilde{\eta}_{(i,t)}$ represent the known cleaning resources requirements of units that are under in-progress offline cleaning at the beginning of the planning horizon of interest.

### 9.3.4   Flexible Time-Window Offline Cleaning Tasks

In general, there may be alternative options for these offline cleaning tasks. And as such, one cleaning task option need to start within the given time window $t = [\tau_i^{\text{es}}, \tau_i^{\text{ls}}]$ as given by

$$\sum_{q \in Q_i} \sum_{t=\tau_i^{\text{es}}}^{\tau_i^{\text{ls}}} H_{(i,q,t)} = 1 \quad \forall i \in \text{FM}_i \tag{9.6}$$

Observe that multiple such cleaning tasks can be modeled for a unit by providing different non-overlapping time windows, if needed.

### 9.3.5   Condition-Based Online Cleaning Tasks

In any given time period, a unit could be under online cleaning only if the unit is under operation during this period, as modeled by

$$V_{(i,t)} \leq X_{(i,t)} \quad \forall i \in CB_i^{on}, t \in T \tag{9.7}$$

In practice, very frequent online cleaning may affect negatively the condition and operation of a unit. For this reason, the proposed approach considers that a unit can undergo an online cleaning task after a minimum time period has passed from the occurrence of the previous online cleaning task in the same unit, as given by

$$\sum_{t'=\max\{1, t-\gamma_i^{on}+1\}}^{t} V_{(i,t')} \leq 1 \quad \forall i \in CB_i^{on}, t \in T$$

$$V_{(i,t)} = 0 \qquad\qquad \forall i \in CB_i^{on}, t \leq (\gamma_i^{on} - \tilde{\gamma}_i^{on}) : \tilde{\gamma}_i^{on} < \gamma_i^{on} \tag{9.8}$$

Parameters $\tilde{\gamma}_i^{on}$ and $\gamma_i^{on}$ represent the total number of time periods that has passed since the last online cleaning at the beginning of the planning horizon and the minimum time between two consecutive online cleaning tasks in a unit, respectively.

$$V_{(i,t)} = \sum_{e \in E_i} V_{(i,e,t)}^{PR} \quad \forall i \in (CB_i^{on} \cap PR_i), t \in T \tag{9.9}$$

Constraints (9.9) relate the two binary variables for online cleaning tasks for the production units. These constraints are needed in order to model correctly the modified maximum operating levels of production units during the period that are under online cleaning. If online cleaning does not affect the maximum operating level of production units, then these constraints can be ignored and variables $V_{(i,e,t)}^{PR}$ do not need to be defined.

### 9.3.6   Condition-Based Cleaning Tasks: Unit Performance Degradation and Recovery

In this study, the performance of any unit that is subject to condition-based maintenance is modeled through the extra energy consumption of the unit $U_{(i,t)}$ due to its deviation from its completely clean condition (i.e., full performance). The performance of the unit decreases as the extra energy consumption increases. To avoid the energy inefficient use and potential damage of the unit, this extra energy consumption for the units under operation should not exceed a maximum extra energy consumption limit $v_i^{max}$, according to

$$U_{(i,t)} \leq v_i^{\max} X_{(i,t)} \quad \forall i \in \mathrm{CB}_i, \forall t \in T \tag{9.10}$$

To continue with, the extra energy consumption of an operating unit is related to: (i) its cumulative time of operation $R_{(i,t)}$, and (ii) its cumulative operating level deviation $D_{(i,t)}$ from its reference operating level (where additional energy consumption is considered minimal), as given by

$$
\begin{aligned}
U_{(i,t)} &\geq \delta_i R_{(i,t)} + \delta_i^q D_{(i,t)} - v_i^{\max}(1 - X_{(i,t)}) \quad \forall i \in \mathrm{CB}_i, \forall t \in T \\
U_{(i,t)} &\leq \delta_i R_{(i,t)} + \delta_i^q D_{(i,t)} + v_i^{\max}(1 - X_{(i,t)}) \quad \forall i \in \mathrm{CB}_i, \forall t \in T
\end{aligned}
\tag{9.11}
$$

Parameters $\delta_i$ and $\delta_i^q$ represent the degradation rates due to the cumulative time of operation and the deviation from the reference operating level, respectively. In industrial applications, it is significant to take into consideration the extra energy consumption contribution due to operation out of the reference operating level since this affects the condition of the equipment. Figure 9.4 presents an illustrative example of two alternative operating level profiles of two units that produce the same product. Observe that the two solutions are equivalent in terms of total production level in any time period. On one hand, the first solution shows many operating level fluctuations and most importantly reports operating levels that are

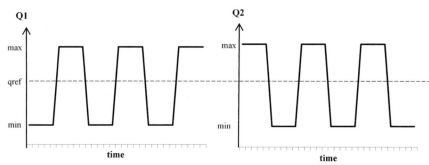

(a) Solution 1: units with different operating levels

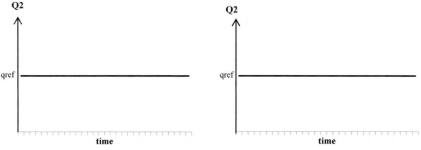

(b) Solution 2: units with same operating levels.

**Fig. 9.4** Illustrative example for operating level deviation of the units

far away from the reference operating level (i.e., this implies additional energy consumption). On the other hand, the second solution reports operating levels for both units equal to the reference operating level in all time periods (i.e., all $D_{(i,t)}$ are zero). In other words, although the two solutions are equivalent in terms of total production, the smooth operation of the second solution results in reduced extra energy consumption and thus slower performance degradation of the unit.

**Cumulative Time of Operation**

The occurrence of an offline cleaning task in a unit resets its cumulative time of operation to zero, according to

$$R_{(i,t)} \leq \bar{\mu}_{(i,t)}(1 - W_{(i,t)}) \quad \forall i \in \mathrm{CB}_i^{\mathrm{off}}, \forall t \in T \tag{9.12}$$

Parameters $\bar{\mu}_{(i,t)}$ are sufficient big numbers. Good values for these parameters for each unit can be calculated through the corresponding maximum extra energy consumption and degradation rate parameters.

The cumulative time of operation for a unit subject to condition-based cleaning is modeled by the following set of constraints:

$$\begin{aligned} R_{(i,t)} &\leq (\tilde{\rho}_i + X_{(i,t)}) + \bar{\mu}_{(i,t)}(W_{(i,t)} + V_{(i,t)}) && \forall i \in \mathrm{CB}_i, \forall t \in T : t = 1 \\ R_{(i,t)} &\leq (R_{(i,t-1)} + X_{(i,t)}) + \bar{\mu}_{(i,t)}(W_{(i,t)} + V_{(i,t)}) && \forall i \in \mathrm{CB}_i, \forall t \in T : t > 1 \end{aligned} \tag{9.13}$$

$$\begin{aligned} R_{(i,t)} &\geq (\tilde{\rho}_i + X_{(i,t)}) - \bar{\mu}_{(i,t)}(W_{(i,t)} + V_{(i,t)}) && \forall i \in \mathrm{CB}_i, \forall t \in T : t = 1 \\ R_{(i,t)} &\geq (R_{(i,t-1)} + X_{(i,t)}) - \bar{\mu}_{(i,t)}(W_{(i,t)} + V_{(i,t)}) && \forall i \in \mathrm{CB}_i, \forall t \in T : t > 1 \end{aligned} \tag{9.14}$$

$$\begin{aligned} R_{(i,t)} &\geq (\tilde{\rho}_i + 1)(1 - \rho_i^{\mathrm{rec}}) - \bar{\mu}_{(i,t)}(1 - V_{(i,t)}) && \forall i \in \mathrm{CB}_i^{\mathrm{on}}, \forall t \in T : t = 1 \\ R_{(i,t)} &\geq (R_{(i,t-1)} + 1)(1 - \rho_i^{\mathrm{rec}}) - \bar{\mu}_{(i,t)}(1 - V_{(i,t)}) && \forall i \in \mathrm{CB}_i^{\mathrm{on}}, \forall t \in T : t > 1 \end{aligned}$$
$$\tag{9.15}$$

For every unit, parameter $\rho_i^{\mathrm{rec}}$ represents the corresponding performance recovery factor due to its online cleaning and parameter $\tilde{\rho}_i$ denotes the cumulative time of operation just before the beginning of the planning horizon of interest (i.e., initial state). Notice that a unit could be subject to both offline and online condition-based cleaning tasks in the proposed approach.

**Cumulative Operating-Level Deviation**

Similarly to the cumulative time of operation, the occurrence of an offline cleaning task in a unit resets its cumulative operating level deviation to zero, according to

$$D_{(i,t)} \leq M_{(i,t)}(1 - W_{(i,t)}) \quad \forall i \in \mathrm{CB}_i^{\mathrm{off}}, \forall t \in T \tag{9.16}$$

Parameters $M_{(i,t)}$ are sufficient big numbers that could be calculated through the corresponding maximum extra energy consumption and degradation rate parameters.

For a utility unit subject to condition-based cleaning, the cumulative operating level deviation from its reference operating level $\left(q_{(i,t)}^{\text{ref}}\right)$ is modeled by the following set of constraints:

$$D_{(i,t)} \leq \tilde{\rho}_i^{dq} + \left(\frac{\left|q_{(i,t)}^{\text{ref}} - \bar{Q}_{(i,t)}\right|}{q_{(i,t)}^{\text{ref}}}\right) + \mu_{(i,t)}(W_{(i,t)} + V_{(i,t)}) + \mu_{(i,t)}(1 - X_{(i,t)})$$

$$\forall i \in (\text{CB}_i \cap \text{UT}_i),\ t \in T : t = 1$$

$$D_{(i,t)} \leq D_{(i,t-1)} + \left(\frac{\left|q_{(i,t)}^{\text{ref}} - \bar{Q}_{(i,t)}\right|}{q_{(i,t)}^{\text{ref}}}\right) + \mu_{(i,t)}(W_{(i,t)} + V_{(i,t)}) + \mu_{(i,t)}(1 - X_{(i,t)})$$

$$\forall i \in (\text{CB}_i \cap \text{UT}_i),\ t \in T : t > 1$$

$$(9.17)$$

$$D_{(i,t)} \geq \tilde{\rho}_i^{dq} + \left(\frac{\left|q_{(i,t)}^{\text{ref}} - \bar{Q}_{(i,t)}\right|}{q_{(i,t)}^{\text{ref}}}\right) - \mu_{(i,t)}(W_{(i,t)} + V_{(i,t)}) - \mu_{(i,t)}(1 - X_{(i,t)})$$

$$\forall i \in (\text{CB}_i \cap \text{UT}_i),\ t \in T : t = 1$$

$$D_{(i,t)} \geq D_{(i,t-1)} + \left(\frac{\left|q_{(i,t)}^{\text{ref}} - \bar{Q}_{(i,t)}\right|}{q_{(i,t)}^{\text{ref}}}\right) - \mu_{(i,t)}(W_{(i,t)} + V_{(i,t)}) - \mu_{(i,t)}(1 - X_{(i,t)})$$

$$\forall i \in (\text{CB}_i \cap \text{UT}_i),\ t \in T : t > 1$$

$$(9.18)$$

$$D_{(i,t)} \geq \left(\tilde{\rho}_i^{dq} + \left(\frac{\left|q_{(i,t)}^{\text{ref}} - \bar{Q}_{(i,t)}\right|}{q_{(i,t)}^{ref}}\right)\right)(1 - \rho_i^{\text{rec}}) - \mu_{(i,t)}(1 - V_{(i,t)})$$

$$\forall i \in (\text{CB}_i^{\text{on}} \cap \text{UT}_i),\ t \in T : t = 1$$

$$(9.19)$$

$$D_{(i,t)} \geq \left(D_{(i,t-1)} + \left(\frac{\left|q_{(i,t)}^{\text{ref}} - \bar{Q}_{(i,t)}\right|}{q_{(i,t)}^{ref}}\right)\right)(1 - \rho_i^{\text{rec}}) - \mu_{(i,t)}(1 - V_{(i,t)})$$

$$\forall i \in (\text{CB}_i^{\text{on}} \cap \text{UT}_i),\ t \in T : t > 1$$

For a production unit subject to condition-based cleaning, the cumulative operating-level deviation from its reference production level $\left(q_{(i,e,t)}^{\text{ref}}\right)$ is modeled by the following set of constraints:

$$D_{(i,t)} \leq \tilde{\rho}_i^q + \left( \frac{\left| q_{(i,e,t)}^{\text{ref}} - Q_{(i,e,t)} \right|}{q_{(i,e,t)}^{\text{ref}}} \right) + \mu_{(i,t)} (W_{(i,t)} + V_{(i,t)}) + \mu_{(i,t)} (1 - X_{(i,t)})$$

$$\forall i \in (\text{CB}_i \cap \text{PR}_i),\ e \in E_i,\ t \in T : t = 1$$

$$D_{(i,t)} \leq D_{(i,t-1)} + \left( \frac{\left| q_{(i,e,t)}^{\text{ref}} - Q_{(i,e,t)} \right|}{q_{(i,e,t)}^{\text{ref}}} \right) + \mu_{(i,t)} (W_{(i,t)} + V_{(i,t)}) + \mu_{(i,t)} (1 - X_{(i,t)})$$

$$\forall i \in (\text{CB}_i \cap \text{PR}_i),\ e \in E_i,\ t \in T : t > 1$$

$$(9.20)$$

$$D_{(i,t)} \geq \tilde{\rho}_i^q + \left( \frac{\left| q_{(i,e,t)}^{\text{ref}} - Q_{(i,e,t)} \right|}{q_{(i,e,t)}^{\text{ref}}} \right) - \mu_{(i,t)} (W_{(i,t)} + V_{(i,t)}) - \mu_{(i,t)} (1 - X_{(i,t)})$$

$$\forall i \in (\text{CB}_i \cap \text{PR}_i),\ e \in E_i,\ t \in T : t = 1$$

$$D_{(i,t)} \geq D_{(i,t-1)} + \left( \frac{\left| q_{(i,e,t)}^{\text{ref}} - Q_{(i,e,t)} \right|}{q_{(i,e,t)}^{\text{ref}}} \right) - \mu_{(i,t)} (W_{(i,t)} + V_{(i,t)}) - \mu_{(i,t)} (1 - X_{(i,t)})$$

$$\forall i \in (\text{CB}_i \cap \text{PR}_i),\ e \in E_i,\ t \in T : t > 1$$

$$(9.21)$$

$$D_{(i,t)} \geq \tilde{\rho}_i^q + \left( \frac{\left| q_{(i,e,t)}^{\text{ref}} - Q_{(i,e,t)} \right|}{q_{(i,e,t)}^{\text{ref}}} \right) (1 - \rho_i^{\text{rec}}) - \mu_{(i,t)} (1 - V_{(i,t)})$$

$$\forall i \in (\text{CB}_i^{\text{on}} \cap \text{PR}_i),\ e \in E_i,\ t \in T : t = 1$$

$$(9.22)$$

$$D_{(i,t)} \geq D_{(i,t-1)} + \left( \frac{\left| q_{(i,e,t)}^{\text{ref}} - Q_{(i,e,t)} \right|}{q_{(i,e,t)}^{\text{ref}}} \right) (1 - \rho_i^{\text{rec}}) - \mu_{(i,t)} (1 - V_{(i,t)})$$

$$\forall i \in (\text{CB}_i^{\text{on}} \cap \text{PR}_i),\ e \in E_i,\ t \in T : t > 1$$

For every unit, parameter $\tilde{\rho}_i^q$ represents its cumulative operating-level deviation just before the beginning of the planning horizon of interest (i.e., initial state).

### 9.3.7   Operational Constraints for Offline Cleaning Tasks

The following set of constraints ensure that a unit that is under offline cleaning remains closed for the whole duration of the selected offline cleaning task option, and relate the two binary variables for offline cleaning tasks

$$X_{(i,t)} + \sum_{t'=\max\{\tau_i^{es}, t-v_{(i,q)}+1\}}^{\min\{\tau_i^{ls}, t\}} H_{(i,q,t')} \leq 1 \tag{9.23}$$

$$\forall i \in (FM_i \cup CB_i^{off}), \ q \in Q_i, \ \tau_i^{es} \leq t \leq (\tau_i^{ls} + v_{(i,q)} - 1)$$

$$W_{(i,t)} = \sum_{q \in Q_i} H_{(i,q,t)} \quad \forall i \in (FM_i \cup CB_i^{off}), \ t \in T : \tau_i^{es} \leq t \leq \tau_i^{ls} \tag{9.24}$$

For condition-based offline cleaning tasks, earliest and latest starting times should be set equal to the first and the last period of the planning horizon, respectively.

### 9.3.8 Resource Constraints for Cleaning Tasks

In the same line with our previous work (Zulkafli and Kopanos 2016), a limited amount of available resources for cleaning operations shared by all types of cleaning tasks is considered, according to

$$\sum_{i \in CB_i^{on}} \vartheta_i^{on} V_{(i,t)} + \sum_{i \in CB_i^{off}} \sum_{q \in Q_i} \sum_{t'=t-v_{(i,q)}+1}^{t} \vartheta_{(i,q)}^{off} H_{(i,q,t')}$$

$$+ \sum_{i \in FM_i} \sum_{q \in Q_i} \sum_{t'=\max\{\tau_i^{es}, t-v_{(i,q)}+1\}}^{\min\{\tau_i^{ls}, t\}} \vartheta_{(i,q)}^{off} H_{(i,q,t')} \leq \eta_t^{max} - \sum_{i \in DM_i} \tilde{\eta}_{(i,t)} \quad \forall t \in T \tag{9.25}$$

For every unit, parameters $\vartheta_i^{on}$ and $\vartheta_{(i,q)}^{off}$ denote the resource requirements for online cleaning and different offline cleaning task options, respectively.

## 9.4 Utility and Product Resources

### 9.4.1 Utility System: Operating Level Bounds

The utility system consists of a number of utility units that could generate the whole set of utility resources required by the production system. If a utility unit operates, its operating level should be between its lower and upper operating level bounds $\left(\kappa_{(i,t)}^{min} \text{ and } \kappa_{(i,t)}^{max}\right)$. Here, changes in the maximum operating levels during online cleaning periods are considered and modeled through: (i) the binary variables related to online cleaning, and (ii) parameters $\pi_i^{on}$ that represent the percentage modification on the upper operating level of a unit that is under online cleaning. Hence, the operating bounds of this general case are given by

$$\kappa_{(i,t)}^{\min} X_{(i,t)} \leq \bar{Q}_{(i,t)} \leq \kappa_{(i,t)}^{\max} (X_{(i,t)} - \pi_i^{\mathrm{on}} V_{(i,t)}) \quad \forall i \in (\mathrm{UT}_i \cap \mathrm{CB}_i^{\mathrm{on}}), \, t \in T \qquad (9.26)$$

Notice that parameters $c_i^{\mathrm{on}}$ are activated only if there is an online cleaning task for a unit. In the case that there is no effect on the maximum operating level of some units during their online cleaning, the corresponding parameters $\pi_i^{\mathrm{on}}$ of these units are set equal to zero. There are some types of utility units, such as combined heat and power units, which generate at the same time more than one utility resources. The generated amount of any utility resource from each utility unit per time period is modeled by

$$Q_{(i,e,t)} = \rho_{(i,e)} \bar{Q}_{(i,t)} \quad \forall i \in UT_i, \, e \in E_i, \, t \in T \qquad (9.27)$$

Parameters $\rho_{(i,e)}$ denote the stoichiometry coefficients that relate the operating level of the utility unit with the generated amount of each utility resource type $(Q_{(i,e,t)})$ that is cogenerated by the same utility system (e.g., heat to power ratio of a combined heat and power unit).

## 9.4.2   Production System: Production Level Bounds

The production system consists of a number of production units that produce the whole set of product resources required by the customers. Here, the production process is modeled as single-stage with a number of units operating in parallel. In order to model the production statuses and levels for production units, the following binary variables are introduced:

$$Y_{(i,e,t)} = \begin{cases} 1 & \text{if production unit } i \in \mathrm{PR}_i \text{ produces product resource } e \text{ in time period } t, \\ 0 & \text{otherwise.} \end{cases}$$

If a production unit produces a product resource $e$, its production level should be between its lower and upper production level bounds ($\bar{\kappa}_{(i,e,t)}^{\max}$ and $\bar{\kappa}_{(i,e,t)}^{\min}$). Similarly to utility units, changes in the maximum production levels during online cleaning periods are considered. Therefore, the production bounds of this general case are given by

$$\bar{\kappa}_{(i,e,t)}^{\min} Y_{(i,e,t)} \leq Q_{(i,e,t)} \leq \bar{\kappa}_{(i,e,t)}^{\max} (Y_{(i,e,t)} - \pi_i^{\mathrm{on}} V_{(i,e,t)}^{\mathrm{PR}}) \quad \forall i \in (\mathrm{PR}_i \cap \mathrm{CB}_i^{\mathrm{on}}), \, e \in E_i, \, t \in T$$
$$(9.28)$$

Online cleaning, as its name implies, could take place in time periods where production units are on operation as modeled by

$$V^{\mathrm{PR}}_{(i,e,t)} \leq Y_{(i,e,t)} \quad \forall i \in (\mathrm{PR}_i \cap \mathrm{CB}^{\mathrm{on}}_i), \, e \in E_i, \, t \in T \qquad (9.29)$$

The two types of operating binary variables for the production units are related by the following set of constraints:

$$\begin{aligned} Y_{(i,e,t)} &\leq X_{(i,t)} && \forall i \in \mathrm{PR}_i, \, e \in E_i, \, t \in T \\ X_{(i,t)} &\leq \sum_{e \in E_i} Y_{(i,e,t)} \leq 1 && \forall i \in \mathrm{PR}_i, \, \forall t \in T \end{aligned} \qquad (9.30)$$

According to these constraints, operating binary variables $X_{(i,t)}$ would be equal to one if and only if there is production of a product resource. In addition, the latter constraints ensure that a production unit could produce at most one product resource per time period.

## 9.4.3   Inventory Tanks

Production and utility systems contain a number of resource-dedicated inventory tanks. These inventory tanks can receive resources $\left(B^+_{(e,z,t)}\right)$ from their associated units $I_z$, according to

$$B^+_{(e,z,t)} = \sum_{i \in (I_e \cap I_z)} Q_{(i,e,t)} \quad \forall e \in E, \, z \in Z_e, \, t \in T \qquad (9.31)$$

Lower and upper bounds on the inlet flows of resources to inventory tanks are considered by:

$$\varepsilon^{+,\min}_{(e,z,t)} \leq B^+_{(e,z,t)} \leq \varepsilon^{+,\max}_{(e,z,t)} \quad \forall e \in E, \, z \in Z_e, \, t \in T \qquad (9.32)$$

Resource balances for every resource-dedicated inventory tank per time period are given by

$$\begin{aligned} B_{(e,z,t)} &= \tilde{B}_{(e,z)} + B^+_{(e,z,t)} - B^-_{(e,z,t)} && \forall e \in E, \, z \in Z_e, \, t \in T : t = 1 \\ B_{(e,z,t)} &= (1 - \beta^{\mathrm{loss}}_z)B_{(e,z,t-1)} + B^+_{(e,z,t)} - B^-_{(e,z,t)} && \forall e \in E, \, z \in Z_e, \, t \in T : t > 1 \end{aligned} \qquad (9.33)$$

Notice that variables $B_{(e,z,t)}$ indicate the inventory level per resource and inventory tank at the end of each time period and variables $B^-_{(e,z,t)}$ represent the outlet resource flow from each inventory tank. Parameters $\tilde{\beta}_{(e,z)}$ stand for the initial inventory for each resource inventory tank at the beginning of the planning horizon (i.e., initial state) and parameters $\beta_z^{\text{loss}}$ provide the losses coefficients for each resource inventory tank. Minimum and maximum inventory levels for the inventory tanks are also considered as

$$\zeta_{(e,z)}^{\min} \leq B_{(e,z,t)} \leq \zeta_{(e,z)}^{\max} \quad \forall e \in E, z \in Z_e, t \in T \tag{9.34}$$

The amount of each utility resource that leaves its dedicated inventory tank and its minimum and outlet flows are given by the following set of constraints:

$$B_{(e,z,t)}^{-} = \sum_{i \in (\text{PR}_i \cap I_z)} B_{(e,z,i,t)}^{\text{UT},-} \quad \forall e \in E^{\text{UT}}, z \in Z_e, t \in T \tag{9.35}$$

$$\varepsilon_{(e,z,t)}^{-,\min} \leq B_{(e,z,t)}^{-} \leq \varepsilon_{(e,z,t)}^{-,\max} \quad \forall e \in E^{\text{UT}}, z \in Z_e, t \in T \tag{9.36}$$

### 9.4.4  Demands for Product Resources

The demands for final products $\left(\zeta_{(e,t)}\right)$ should be satisfied for every time period, according to

$$\text{NS}_{(e,t)}^{\text{FP}} + \sum_{z \in Z_e} B_{(e,z,t)}^{-} = \zeta_{(e,t)} \quad \forall e \in E^{\text{PR}}, t \in T \tag{9.37}$$

Variables $\text{NS}_{(e,t)}^{\text{FP}}$ denote the amount of the demand for each product resource $(E^{\text{PR}})$ per time period that cannot be satisfied by the internal production system. These unsatisfied demands for product resources should be covered by acquiring product resources from external sources. Generally speaking, this is highly undesirable and for this reason, a very high penalty or purchase cost is usually used in the optimization goal. If product resources cannot be acquired from external sources, variables $\text{NS}_{(e,t)}^{\text{FP}}$ present the lost sales of product resources.

### 9.4.5  Demands for Utility Resources (Link Between Utility and Production Systems)

The requirements for utility resources give the linking constraints between utility and production systems. For each time period, the demands for utility resources per production unit $I_e^{PR}$ consist of: (i) fixed utility resource requirements that depend on the operational status of the production unit; and (ii) variable utility resource requirements that depend on the production level of the production unit

$$NS_{(e,i,t)}^{UT} + \sum_{z \in (Z_e \cap Z_i)} B_{(e,z,i,t)}^{UT,-} = \sum_{e' \in (E^{PR} \cap E_i)} (\alpha_{(i,e,e')} Q_{(i,e',t)} + \bar{\alpha}_{(i,e,e')} Y_{(i,e',t)})$$

$$\forall e \in E^{UT}, i \in I_e^{PR}, t \in T$$

(9.38)

Variables $NS_{(e,i,t)}^{UT}$ represent the amount of unsatisfied demand for each utility resource per time period. Similar to the unsatisfied demand for product resources, penalty or purchase costs for acquiring utility resources from external sources are typically introduced in the objective function of the optimization problem.

## 9.5  Objective Function

The optimization goal is to minimize the total cost of the production and the utility system. More specifically, the objective function includes: (i) startup and shutdown costs for units that are subject to startup and shutdown actions; (ii) variable and fixed operating costs for utility units; (iii) variable and fixed production costs for production units; (iv) penalty or purchase costs for acquiring product and utility resources from external sources; (v) total extra energy consumption costs for utility and production units that are subject to performance degradation modeling; and (vi) total cleaning costs related to online and offline cleaning tasks of production and utility units that are subject to performance degradation. The optimization goal is given by

$$\min \begin{bmatrix} \sum_{t \in T} \sum_{i \in I^{SF}} (\phi_{(i,t)}^{S} S_{(i,t)} + \phi_{(i,t)}^{F} F_{(i,t)}) + \sum_{t \in T} \sum_{i \in I^{UT}} (\phi_{(i,t)}^{UT,op-var} \bar{Q}_{(i,t)} + \phi_{(i,t)}^{UT,op-fix} X_{(i,t)}) \\ + \sum_{t \in T} \sum_{i \in PR_i} \sum_{e \in E_i} (\phi_{(i,e,t)}^{PR,op-var} Q_{(i,e,t)} + \phi_{(i,e,t)}^{PR,op-fix} Y_{(i,e,t)}) \\ + \sum_{t \in T} \sum_{e \in E^{PR}} \phi_{(e,t)}^{PR,ex} NS_{(e,t)}^{FP} + \sum_{t \in T} \sum_{e \in E^{UT}} \sum_{i \in I_{PR}^e} \phi_{(e,i,t)}^{UT,ex} NS_{(e,i,t)}^{UT} \\ + \sum_{t \in T} \sum_{i \in CB_i} \phi_{(i,t)}^{pw} U_{(i,t)} + \sum_{t \in T} \left( \sum_{i \in CB_i^{on}} \phi_{(i,t)}^{on} V_{(i,t)} + \sum_{i \in (CB_i^{off} \cup FM_i)} \sum_{q \in Q_i} \phi_{(i,q,t)}^{off} H_{(i,q,t)} \right) \end{bmatrix}$$

(9.39)

In the above expression, the small-letter symbols correspond to the associated cost coefficients of the corresponding optimization variables. A detailed definition of them can be found in the Nomenclature.

## 9.6   Remarks on Rolling Horizon

Terminal constraints should be defined for some key optimization variables when a rolling horizon approach is used. These constraints are applied for the last time period $|T|$ of the considered prediction horizon and can be typically related to desired minimum resource inventory levels or unit performance levels, as modeled below:

$$
\begin{aligned}
B_{(e,z,t)} &\geq \lambda^B_{(e,z)} \varsigma^{\max}_{(e,z)} \quad \forall e \in E, z \in Z_e, t \in T : t = |T| \\
U_{(i,t)} &\leq \lambda^U_i v^{\max}_i \qquad \forall i \in CB_i, t \in T : t = |T|
\end{aligned}
\tag{9.40}
$$

Parameters $\lambda^B_{(e,z)}$ and $\lambda^U_i$ represent are percentage coefficients used to determine the minimum inventory level for each resource and the maximum extra energy consumption level for each operating unit at the last period of each prediction horizon. In the same line, terminal constraints could be defined for other variables if needed. Generally speaking, terminal constraints are defined as a mean of preserving the stability of the system over its long-term operational horizon. It is also usual to apply terminal constraint values even in deterministic optimization approaches, in order to ensure a better state of the system at the end of the planning horizon. More details about rolling horizon approaches can be found in Kopanos and Pistikopoulos (2014).

## 9.7   Case Studies

In this part, three case studies for the integrated planning problem of utility and production systems are presented in order to highlight the special features of the proposed optimization framework. More specifically, the first case study studies only a flexible time-window cleaning policy for units while the second case study considers both flexible time-window and condition-based cleaning policies for production and utility units. The third case study deals with the reactive planning problem under a rolling horizon approach and considers condition-based cleaning policies for all units. All case studies have been solved following both the proposed integrated approach and the traditional sequential approach. Detailed comparisons between the solutions of both approaches have been made. All resulting optimization problems have been solved in GAMS/CPLEX in an Intel(R) Core (TM) i7 under standard configurations and a zero optimality gap.

### 9.7.1 Case Study 1: Integrated Planning of Utility and Production Systems (Flexible Time-Window Cleaning)

In this case study, flexible time-window offline cleaning tasks for utility and production units are only considered (i.e., no condition-based maintenance). All parameters are deterministic.

### 9.7.2 Description of Case Study 1

The system under consideration consists of five utility units $(i1-i5)$ and three production units $(i6-i8)$. The utility units can produce two utility resources $(e1, e2)$, which could be either stored in their associated inventory tanks $(z1, z2)$ or consumed directly by the production units. Two final product resources $(e3, e4)$ can be produced by the production units that can be either stored in their dedicated inventory tanks $(z3, z4)$ or meet directly the customer demand. Each utility and production unit has a maximum operating level, as given by Table 9.1. Minimum operating levels for units are 10% of the corresponding maximum operating levels. For each production unit and product resource, Table 9.2 provides the stoichiometric coefficients of fixed and varied utility needs for the production of a unit of the associated product resource. Table 9.3 gives the cogeneration coefficient of each utility resource for every utility units. For example, for utility unit $i1$, four units of $e2$ are generated for every unit of $e1$ produced. Notice that utility unit $i4$ and $i5$ cannot generate utility resource $e2$ and $e1$, respectively. Maximum runtimes for units are not considered. There is a maximum number of available resources for

**Table 9.1** Case study 1: maximum operating levels for utility and production units

| $\kappa^{max}_{(i,e,t)}$ | $i1$ | $i2$ | $i3$ | $i4$ | $i5$ | $i6$ | $i7$ | $i8$ |
|---|---|---|---|---|---|---|---|---|
| $e1$ | 50 | 80 | 60 | 60 | – | – | – | – |
| $e2$ | 200 | 160 | 180 | – | 140 | – | – | – |
| $e3$ | – | – | – | – | – | 85 | 65 | 50 |
| $e4$ | – | – | – | – | – | 65 | 50 | 85 |

**Table 9.2** Case study 1: fixed and varied stoichiometric coefficients of utility needs for production units (per unit of product resource)

| Unit | Product | $\alpha_{(i,e,e3)}$ | $\alpha_{(i,e,e4)}$ | $\bar{\alpha}_{(i,e,e3)}$ | $\bar{\alpha}_{(i,e,e4)}$ |
|---|---|---|---|---|---|
| $i6$ | $e1$ | 0.90 | 0.80 | 17 | 15 |
|  | $e2$ | 2.25 | 3.38 | 45 | 39 |
| $i7$ | $e1$ | 0.80 | 0.70 | 14 | 18 |
|  | $e2$ | 3.38 | 5.25 | 54 | 30 |
| $i8$ | $e1$ | 0.75 | 0.90 | 16 | 10 |
|  | $e2$ | 2.63 | 3.00 | 36 | 48 |

**Table 9.3** Case study 1: cogeneration coefficients of utility units per utility resource

| $P_{(i,e)}$ | $e1$ | $e2$ |
|---|---|---|
| $i1$ | 1 | 4 |
| $i2$ | 1 | 2 |
| $i3$ | 1 | 3 |
| $i4$ | 1 | 0 |
| $i5$ | 0 | 1 |

cleaning tasks equal to 12 cleaning resource units. The minimum runtime for utility and production units $(\omega_i)$ is 6 days and the minimum offline time after shutdown $(\psi_i)$ is 3 days. No lower bounds are considered for minimum inventory level $\left(\zeta_{(e,z)}^{min}\right)$, minimum flows of resources to inventory tanks $\left(\varepsilon_{(e,z,t)}^{+,min}\right)$ and minimum flows of resources leaves inventory tanks $\left(\varepsilon_{(e,z,t)}^{-,min}\right)$. There is no maximum resources flow constraint to inventory tank $\left(\varepsilon_{(e,z,t)}^{+,max}\right)$. The maximum inventory level $\left(\zeta_{(e,z)}^{max}\right)$ for resources $e1, e2, e3$, and $e4$ are 100, 320, 200, and 300 units, respectively. The maximum flows of utility resources leaving their respective inventory tank $\left(\varepsilon_{(e,z,t)}^{-,max}\right)$ are 400 units for utility resource $e1$ and 600 units for utility resource $e2$.

A total planning horizon of 30 days, divided into daytime periods (i.e., 30 time periods), is considered. All utility and production units should undergo a flexible time-window offline cleaning task. The earliest/latest cleaning startup times $\left(\tau_i^{es}/\tau_i^{ls}\right)$ are on day 9 and 15 for utility units and on day 20 and 25 for production units, respectively. There are three alternative flexible time-window offline cleaning options $(q1, q2, q3)$ that are characterized by different durations, cleaning resources requirements and associated costs, as shown in Table 9.4. Operational costs for utility and production units are given in Table 9.5. Purchase costs for utility and product resources are 6000 and 4000 m.u./unit, respectively.

The initial inventory for resources $e1, e2, e3$ and $e4$ is 10, 20, 50, and 300 units, respectively. It is assumed that the process plant is closed before the beginning of the planning horizon of interest, therefore there is no initial state (i.e., $\tilde{\chi}_i, \tilde{\psi}_i,$ or $\tilde{\omega}_i$) that is taken into account for this case study. In addition, Fig. 9.5 shows the normalized demand for product resources by having the peak demand value of

**Table 9.4** Case study 1: alternative options for flexible time-window offline cleaning tasks

| Units | Parameter | Metric unit | $q1$ | $q2$ | $q3$ |
|---|---|---|---|---|---|
| $i1-i8$ | $v_{(i,q)}$ | Days | 3 | 4 | 5 |
| $i1-i8$ | $\vartheta_{(i,q)}^{off}$ | Resource units | 6 | 4 | 3 |
| $i1, i2, i5-i8$ | $\phi_{(i,q,t)}^{off}$ | m.u./cleaning | 2137.5 | 1425.0 | 1068.8 |
| $i3$ and $i4$ | $\phi_{(i,q,t)}^{off}$ | m.u./cleaning | 7087.5 | 4725.0 | 3543.8 |

**Table 9.5** Case study 1: operational costs for utility and production units

| Units | Resource | $\phi^{S}_{(i,t)}$ (m.u./unit) | $\phi^{F}_{(i,t)}$ (m.u./unit) | $\phi^{fix}_{(i,e,t)}$ (m.u./unit) | $\phi^{var}_{(i,e,t)}$ (m.u./unit) |
|-------|----------|---------|---------|-----------|-----------|
| i1 | e1 and e2 | 2300 | 1150 | 220 | 10 |
| i2 | e1 and e2 | 2350 | 1170 | 250 | 10 |
| i3 | e1 and e2 | 2370 | 1200 | 270 | 10 |
| i4 | e1 | 2250 | 1000 | 150 | 15 |
| i5 | e2 | 2270 | 1050 | 200 | 15 |
| i6 | e3 \| e4 | 2300 | 1150 | 500 \| 400 | 1.2 \| 1.0 |
| i7 | e3 \| e4 | 2000 | 1100 | 400 \| 300 | 1.5 \| 1.4 |
| i8 | e3 \| e4 | 2300 | 1150 | 300 \| 500 | 1.4 \| 1.9 |

product resource e4 as a reference. The range for the demand for product resource e3 is between 40 and 100 units and for product resource e4 is between 50 and 120 unit, respectively.

### 9.7.3   Results of Case Study 1—Integrated Approach

This example has been solved by using the proposed integrated optimization framework, and the results obtained are reported, analyzed and discussed below.

Figure 9.6 displays the optimal operational and cleaning plan for both the utility and the production system. More specifically, this figure shows for each unit per time period: (i) the operational status (i.e., in operation, idle, startup, shutdown, or under cleaning), (ii) the selected offline cleaning task options, (iii) the type of utility or product resources produced from each unit, and (iv) the profile of the cleaning resources requirements. No performance level profiles are displayed in this case study because no condition-based cleaning tasks are considered here.

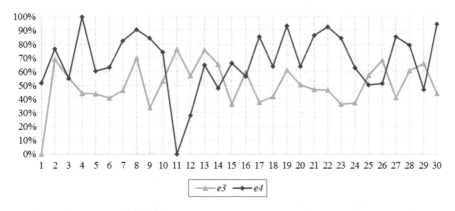

**Fig. 9.5** Case study 1: normalized demand profiles for products per time period

Simultaneous cleaning tasks between utility units are observed. For instance, utility units $i4$ and $i5$ are under cleaning from day 9 to 11 and utility units $i2$ and $i3$ are under cleaning from day 12 to 14. In addition, it is observed a simultaneous cleaning for utility unit $i1$ and production unit $i8$ from day 15 to 17. The flexible time-window for the cleaning of production units is long enough to avoid simultaneous cleaning tasks of multiple production units. Notice that in the optimal solution the most expensive cleaning option $q1$ (but with the smaller duration) has only been selected most probably because of: (i) the overall high demands for product resources throughout the planning horizon of interest; (ii) the relatively narrow flexible time windows for the cleaning of utility units; (iii) the constrained availability of cleaning resources per time period; and (iv) the high purchase costs for utility and product resources.

Utility unit $i4$, which can generate only utility resource $e1$, is not operating in day 1 and day 8, because there is enough supply of utility $e1$ from the other utility units and its corresponding inventory tank. Production unit $i7$ is idle from day 9 to 14 mainly due to following two reasons: (i) two utility units are under cleaning during these periods (see Fig. 9.6) a fact that decreases the total utility generation capacity of the plant and therefore the total production capacity as well; and (ii) the total demands for products are relatively lower in these time periods (see Fig. 9.5).

Figure 9.7 displays the normalized operating level profiles for all utility and production units. The maximum operating level of each unit has been used as a reference of normalization (see Table 9.1). In the utility system, utility units $i1$ to $i3$ operate at their maximum operating levels throughout the planning horizon (excluding their cleaning periods). It is observed that utility unit $i4$ that can generate only utility $e1$ and utility unit $i5$ that can generate only utility $e2$ operate in a

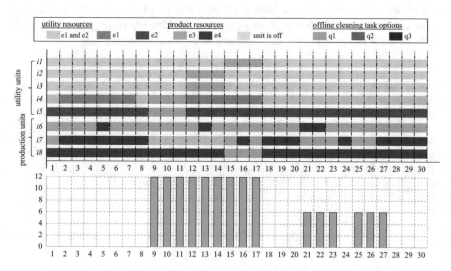

**Fig. 9.6** Case study 1—integrated Approach: Optimal operational and cleaning plan for production and utility systems and total utilization profile of cleaning resources

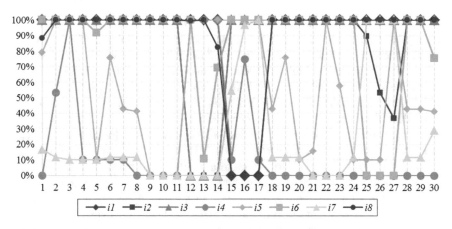

**Fig. 9.7** Case study 1—integrated approach: normalized operating level profiles for utility and production units

broader operating range to cover the fluctuations of the utility requirements of the production system. In the production system, production units $i6$ and $i8$ operate at their maximum capacities most of the time periods, while production unit $i7$ operates at its minimum capacity. The latter is observed basically due to the relatively high shutdown costs compared to fixed and variable operating cost at the minimum operating level. For this reason, it is preferred to continue operating this production unit at minimum capacity and avoid shutting it down, since this would impose a considerable shutdown cost.

Figure 9.8 displays the normalized total production profiles for every utility and final product resource. The production of each resource is calculated by having the cumulative production of the resource from each unit divided by the maximum total

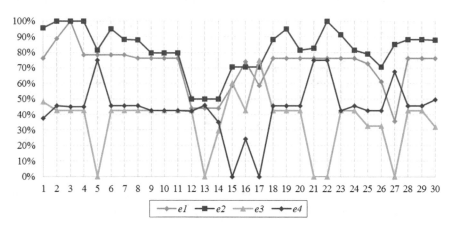

**Fig. 9.8** Case study 1—integrated approach: normalized total production profiles for utility and final product resources

resource capacity of all units. Not surprisingly, it is observed that the trend of the total production profile for e3 follows the opposite trend of that of e4, since the limited number of production units can produce at most one final product per time period. For instance, the high total production peak levels for product resource e4 instead of low total production levels for product e3 in days 5, 13, 21, 22, and 27 are due to the fact that the production units produce exclusively product e4 in all these days (see also Fig. 9.6). The opposite trend is observed on day 15, and 17 when high total peak levels for product e3 but low levels for product e4 when production units produce only product e3 in these days. Meanwhile, the production trends for utilities e1 and e2 follow quite a similar trend throughout the planning horizon, mainly due to the presence of three utility units that cogenerate both utility resources. For example, there is a reduction in the total operating levels for utility resources e1 and e2 when the utility units undergo cleaning between day 9 and 15.

Figure 9.9 displays the normalized inventory profiles for utility and product resources, having as reference the corresponding maximum inventory level of each inventory tank. Low utility inventory levels from day 12 to 20 are mainly due to reduced utility capacities, because utility units i1, i2, and i3 are under cleaning tasks in this period (see Fig. 9.6). Importantly, there is no purchase of utility or product resources at any time period. From day 20 and onwards, the inventory levels of product resource e3 are low because of: (i) the occurrence of a cleaning task in production unit i6 (see Fig. 9.6); and (ii) its high demands (as shown in Fig. 9.5). Similarly, the low inventory profile for product e4 from day 17 and onwards is due to its higher demand and the cleaning of production unit i7 started on day 21.

Figure 9.10 shows the breakdown of the total cost for the utility and the production systems. The costs are divided into: (i) the startup and shutdown operations; (ii) the operation of the utility system; (iii) the operation of the production system; (iv) the offline cleaning tasks for the units; and (v) the total purchase of utility and product resources. The operational cost for the utility system remains the highest cost term at about 46% of the total cost. The second highest cost is the startup and

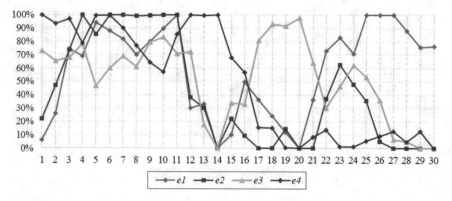

**Fig. 9.9** Case study 1—integrated approach: normalized inventory profiles for utility and product resources

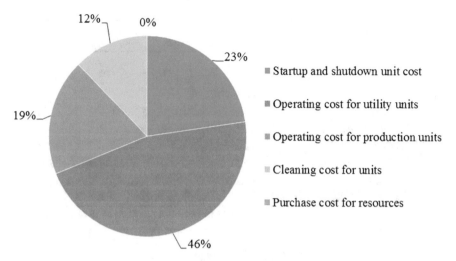

**Fig. 9.10** Case study 1—integrated approach: total cost breakdown (percentage)

shutdown units costs, which is about 23% of the total cost, because of the initial state of the overall system (the plant was closed before the beginning of the planning horizon). The cleaning cost is around 12% of the total cost while there is no purchase cost.

## 9.7.4   Results of Case Study 1—Sequential Approach

Here, the same case study has been solved considering the traditional sequential approach, where the planning problem of the production system is solved first using simply upper bounds on the total available utility amounts per time period. After the solution of this production planning problem, the associated variables that describe the production of final products (i.e., $Q_{(i,t)}$ and $Y_{(i,e,t)}$), product inventories and flows (i.e., $B_{(e,z,t)}$, $B^-_{(e,z,t)}$, and $B^+_{(e,z,t)}$) and occurrence of cleaning tasks in the production units (i.e., $H_{(i,q,t)}$) are fixed, and the planning problem of the utility system is solved.

Figure 9.11 displays the operational and cleaning plan for the production and the utility system obtained by following the sequential approach. In this case, cleaning tasks options $q2$ and $q3$ are selected for the production units. It should be emphasized, in contrast to the solution of the integrated approach, the solution of the sequential approach reports purchases of utilities from external sources in some time periods, as shown in Fig. 9.12. In particular, important utility purchases are observed between day 10 and 16 because of the occurrence of multiple cleaning tasks in the utility units over this time window (see Fig. 9.12). Furthermore, utility units $i4$ and $i5$ operate in less time periods in the solution of the sequential approach

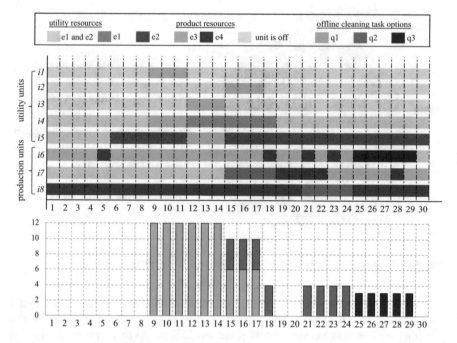

**Fig. 9.11** Case study 1—sequential approach: operational and cleaning plan for production and utility systems and total utilization profile of cleaning resources

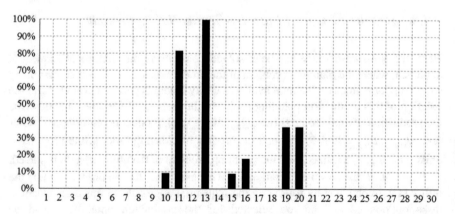

**Fig. 9.12** Case study 1— sequential approach. Normalized profile of total purchases for utilities

than in that of the integrated approach which causes the need for utility purchases (see Fig. 9.6). A total of 633 units of utility resources need to be purchased throughout the planning horizon. If there is no option of acquiring utilities from externals sources, this would make the production plan infeasible in practice. The total cost of the solution following the integrated approach is more than 5% lower than that of the solution found by the sequential approach, which is a clear evidence of the benefits that the proposed integrated approach can have over its sequential counterpart.

## 9.8 Case Study 2: Integrated Planning of Utility and Production System (Condition-Based Cleaning and Flexible Time-Window Cleaning)

In this case study, a condition-based cleaning policy for utility units and a flexible time-window cleaning policy for production units are considered. The condition-based cleaning policy involves online and offline cleaning tasks. All parameters are deterministic.

### 9.8.1 Description of Case Study 2

Here a modified version of the previous case study is considered. The main parameters (Tables 9.1, 9.2, 9.3 and 9.4) and operational costs (Table 9.5) are the same as in Case Study 1. Minimum runtime and shutdown times are the same as in Case Study 1. The demand for products for this case study follows the same pattern as in the previous example, but reduced by 15%. A main difference here is that the utility units $(i1-i5)$ should undergo condition-based cleaning tasks. Meanwhile, production unit $i6$ has a fixed offline cleaning and the other production units $(i7-i8)$ should undergo flexible time-window offline cleaning tasks. The earliest and latest cleaning startup times $(\tau_i^{es}/\tau_i^{ls})$ for production units $i7$ and $i8$ are in day 15 and 25, respectively. As before, there are three alternative cleaning tasks options that can be selected for condition-based offline cleaning (i.e., utility units) and time-window flexible cleaning (i.e., production units). The maximum available resources per time period for the cleaning tasks are 12 units of cleaning resources. The parameters that refer to condition-based offline and online cleaning for utility units are defined as follows: (i) the extra power consumption limit $(v_i^{max})$; (ii) performance degradation rate $(\delta_i)$; (iii) performance coefficient related to operating level $(\delta_i^q)$; (iv) minimum time between two consecutive online cleaning tasks $(\gamma_i^{on})$; (iv) the recovery factor of the online cleaning for any utility unit $(\rho_i^{rec})$; (v) references operating level $(q_{(i,t)}^{ref})$; and (iv) the resource requirement of online cleaning $(\vartheta_i^{on})$ as shown in Table 9.6.

At the end of the planning horizon of interest, there are two types of terminal constraints for the: (i) inventory levels of utility and product resources; and (ii) the performance level of the operating utility units. Namely, at the end of the planning horizon, the inventory levels of each resource should be greater or equal to 25%

**Table 9.6** Case study 2: parameters related to the condition-based cleaning of utility units

| Parameter | i1 | i2 | i3 | i4 | i5 |
|---|---|---|---|---|---|
| $v_i^{max}$ | 162 | 153 | 247 | 200 | 210 |
| $\delta_i$ | 9 | 9 | 13 | 10 | 10 |
| $\delta_i^q$ | 6.75 | 6.75 | 9.75 | 7.50 | 7.50 |
| $\gamma_i^{on}$ | 10 | 10 | 10 | 10 | 10 |
| $\rho_i^{rec}$ | 0.20 | 0.20 | 0.20 | 0.20 | 0.20 |
| $q_{(i,t)}^{ref}$ | 50 | 80 | 60 | 60 | 70 |
| $\vartheta_i^{on}$ | 1 | 1 | 1 | 1 | 1 |

**Table 9.7** Case study 2: initial state of the utility and production system

| Parameter | i1 | i2 | i3 | i5 |
|---|---|---|---|---|
| $\tilde{\rho}_i$ | 2 | 4 | 2 | 2 |
| $\tilde{\beta}_{(e1,z1)}$ | 10 | Units | Initial inventory for utility $e1$ | |
| $\tilde{\beta}_{(e2,z2)}$ | 20 | Units | Initial inventory for utility $e2$ | |
| $\tilde{\beta}_{(e3,z3)}$ | 50 | Units | Initial inventory for product $e3$ | |
| $\tilde{\beta}_{(e4,z4)}$ | 300 | Units | Initial inventory for product $e4$ | |

from its corresponding maximum inventory level $\left(\zeta_{(e,z)}^{max}\right)$, and the performance level of each utility unit that is under operation at the end of the planning should be greater or equal to 25% (i.e., lower or equal to 75% of the corresponding $v_i^{max}$). In addition, Table 9.7 gives the values of the parameters that define the initial state of the utility and production systems. All other initial state parameters are zero.

## 9.8.2 Results of Case Study 2—Integrated Approach

Figure 9.13 displays the optimal operational and cleaning plan for both production and utility system. For each production and utility unit: (i) the operational status at each time period; (ii) the selected offline cleaning tasks options and online cleaning tasks on its corresponding time period; (iii) the type of utility or product resources produced from each unit; and (iv) the profile of the cleaning resources requirements are observed.

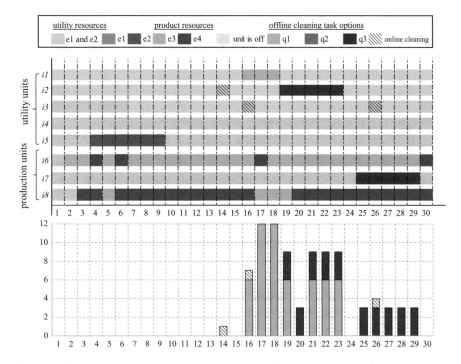

**Fig. 9.13** Case study 2—integrated approach: optimal operational and cleaning plan for production and utility systems and total cleaning resources utilization profile

Simultaneous condition-based offline cleaning tasks are observed for utility unit $i1$ and production unit $i8$ in day 17 and 18. The solution reports condition-based cleaning tasks for utility units $i1$ to $i3$. Meanwhile, utility unit $i4$ that can only produce utility $e1$ remains closed for all time periods because utility resource $e1$ has enough supply from other utility units (e.g., $i1, i2$ and $i3$) that can cogenerate both utility resources. Utility unit $i5$ which can only produce utility resource $e2$ operates in a shorter duration from day 4 to 9 because utility unit $i3$ is closed. The demand for utility resource $e2$ cannot be satisfied by just utility unit $i1$ and $i2$, thus utility unit $i5$ operates to fully satisfy this demand in these days. Production unit $i6$ produces product resource $e3$ and production unit $i8$ produces product resource $e4$ in most of the time periods. This should be due to the stoichiometric coefficient $\alpha_{(i,ee,e)}$ and $\beta_{(i,ee,e)}$ that define the utility requirements per product unit (see Table 9.2). Another observation is that production unit $i7$ remains idle throughout planning horizon but there is a predefined flexible cleaning task option $q3$ that starts in day 25. It should clear here that the longest duration cleaning task option is selected due to its lower cost. In reality, the production manager may find that this cleaning is not necessary because this production unit does not operate within the current planning horizon, and may ignore it.

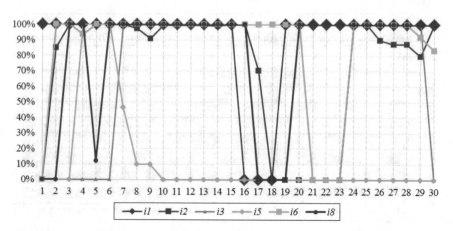

**Fig. 9.14** Case study 2—integrated approach: normalized operating level profiles for utility and production units

Figure 9.14 displays the normalized operating level profiles for utility and production units, having as a reference the maximum operating level of each unit as given in Table 9.1. In the utility system, utility units $i1$ to $i3$ operate at their maximum operating levels throughout the planning horizon (excluding their cleaning periods). Utility unit $i5$, which can generate only utility resource $e2$, operates from day 4 to 9 to satisfy the needs for utility resource $e2$. Maximum production level for utility units $i5$ is observed from day 4 to 6 because utility unit $i3$ is offline (refer to Fig. 9.13). Then, the production level for utility unit $i5$ reduces to minimum because utility unit $i3$ starts up on day 7. In the production system, production units $i6$ and $i8$ operate at their maximum capacity almost in all time periods in order to satisfy the high demand for product resources.

Figure 9.15 displays the normalized total production profiles for every utility and final product resource. The total production for each resource is calculated by having the cumulative production of the resource from each unit divided by the maximum total resource capacity from all units. The production trends for utility resources $e1$ and $e2$ follow quite a similar trend throughout the planning horizon, mainly due to the presence of three utility units that cogenerate both utility resources. The only differences are observed when utility unit $i5$ operates from day 4 to 9. There are higher production differences for utility resource $e2$ than that of the production of utility resource $e1$. The total production level for utility resources $e1$ and $e2$ are considerably reduced when cleaning takes place for utility units between days 16 and 23. The production profiles for product resources $e3$ and $e4$ from day 7 to 14 and from day 24 to 28 are on the same level because the upper operating level of utility unit $i6$ (produces product resource $e3$) and utility unit $i8$ (produces product resource $e4$) in all these days are the same (see Table 9.1). In addition, when there is no production of a product resource in certain time periods (e.g., days 1, 4, 15, 17, 21, 22, 23, 30 for product resource $e3$ and days 1, 2, 5, 18, 19 for product

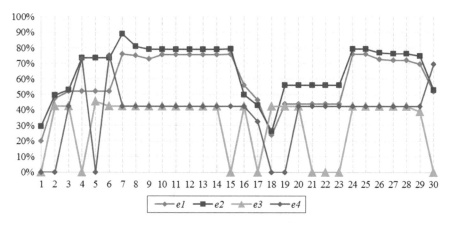

**Fig. 9.15** Case study 2—integrated approach. Normalized total production profiles for utility and final product resources

resource $e4$) its corresponding demand is fully satisfied with its associated inventory tank.

Figure 9.16 displays the normalized inventory profiles for utility and product resources. The maximum inventory levels $\left(\zeta_{(e,z)}^{max}\right)$ are the reference values here. It is observed that, high inventory level for utility and product resources at the beginning of planning horizon because of initial inventory levels. There are reduced inventory levels for utility and product resources on day 16–23 because cleaning of utility unit $i1$ and $i2$ and production unit $i6$ and $i8$ take place on these days. At the end of day 30, the inventory level for utility $e2$ and product $e3$ and $e4$ are not approaching zero due to terminal constraints are set to be 25% of the initial

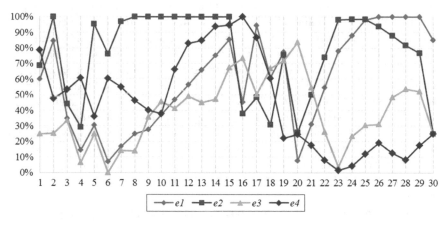

**Fig. 9.16** Case study 2—integrated approach. Normalized inventory profiles for utility and product resources

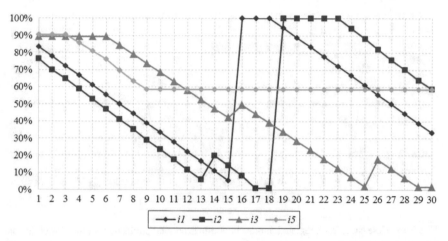

**Fig. 9.17** Case study 2—integrated approach: performance level profiles for utility units per time period

inventory. However, this is not the case for utility $e1$ because all utility units (i.e., $i1$, $i2$ and $i3$) that cogenerate both utilities are operating at their maximum operating capacities (refer to Fig. 9.14). It is not possible to operate these utility units in a lower capacity at the end of the planning horizon because the utility demand for $e2$ must be fully satisfied in order to meet the demand for products. Thus, the optimal solution reports a 25% of inventory level for utility $e2$ and a much higher inventory level for utility $e1$ at the end of the planning horizon.

Figure 9.17 shows the performance level profiles for utility units that are subject to condition-based cleaning. The performance level of a unit depends on its cumulative time of operation and its operating levels deviation. Here, it can be seen when the performance of utility units $i1$ and $i2$ is fully recovered once an offline cleaning occurs. It is also observed that utility unit $i2$ partially recovers its performance through an online cleaning on day 14, and it continues operating until reaching its critical performance level on day 17. The performance degradation of utility unit $i5$ declines in a slightly varied rate (i.e., no straight line decline) from day 7 to 9 due to the deviation of its operating level from its maximum operating capacity (see Fig. 9.14). Utility unit $i5$ shuts down in day 10 and remains idle for the remaining planning horizon, thus no cleaning task is performed after its shutdown. The performance levels of all operating utility units at the end of the planning horizon remain above 25% (due to the terminal constraints imposed) except for utility unit $i3$ that does not operate in day 30 and therefore terminal constraint was not applied (see Fig. 9.13). In practice, one could start an offline cleaning task on this unit during the last period of the planning horizon to completely restore its performance.

Figure 9.18 demonstrates the total cost breakdown for the utility and production systems. As in the previous case study, the operating cost of the utility system remains the highest cost term. This is because the production levels of utility

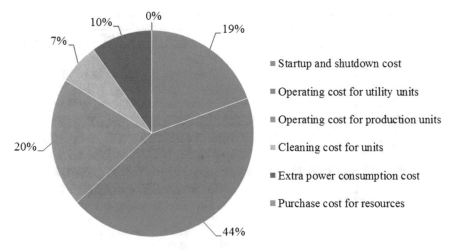

**Fig. 9.18** Case study 2—integrated approach: total cost breakdown (percentage)

resources to satisfy the utility demand of the production system are much higher than the production levels of the production system. Also, variable and fixed utility costs are relatively expensive. The startup and shutdown cost and the operating cost of the production system are at 19 and 20% of the total cost, respectively. The extra energy consumption and cleaning costs are around 10 and 7%.

## 9.8.3   Results of Case Study 2—Sequential Approach

The same case study has been solved following the traditional sequential approach in order to make a comparison of its solution with the solution obtained by the proposed integrated approach. Figure 9.19 displays the optimal operational and cleaning plan for the sequential approach. In comparison with the integrated approach, a higher number of online cleaning tasks for utility units is observed. Some major observations are that: (i) utility unit $i4$ still remains inactive throughout the whole planning horizon; (ii) utility unit $i5$ operates in a larger number of time periods than before; and (iii) production unit $i7$ now operates in most of the time periods and production unit $i8$ operates less time in the 30-day planning horizon.

Figure 9.20 shows the normalized operating level profiles for utility and production units of the solution of the sequential approach. In comparison with the solution of the integrated approach (Fig. 9.14), utility units $i1$ and $i3$ operate at their maximum operating levels while the operating level of utility unit $i2$ varies in order to accommodate the demand for utility resources. Utilized production units operate on their maximum operating capacities most of the times.

Figure 9.21 displays the normalized total production profiles for utility and product resources. The production profiles for utility resources $e1$ and $e2$ follow

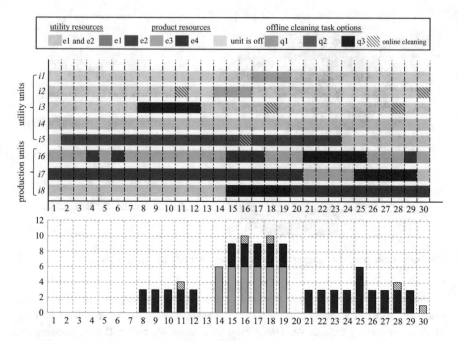

**Fig. 9.19** Case study 2—sequential approach: operational and cleaning plan for production and utility systems and total utilization profile of cleaning resources

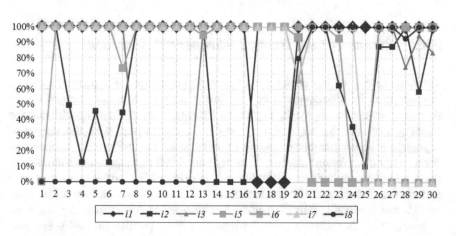

**Fig. 9.20** Case study 2—sequential approach: normalized operating level profiles for utility and production units

quite a similar pattern throughout the planning horizon. Since a production unit can produce at most one product resource per time period and there is a limited number of production units, the production profile for product resource $e3$ follows the opposite trend of that of product resource $e4$.

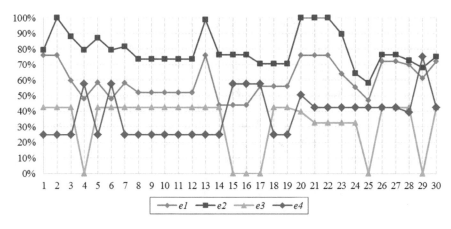

**Fig. 9.21** Case study 2—sequential approach. Normalized total production profiles for utility and product resources

The normalized inventory profiles for utility and product resources are shown in Fig. 9.22. The inventory levels for utility resources $e1$ and $e2$ are lower in day 14–19, which is due to the offline and online cleaning of the utility units (see Fig. 9.19). The inventory level for product resource $e3$ reduces considerably from day 15 and 17 because no production unit is producing product resource $e3$ in these days and the corresponding demand is satisfied exclusively from its inventory tank. At the end of day 30, the inventory level for utility resource $e2$ and product resources $e3$ and $e4$ are equal to 25% of their maximum inventory capacity due to the terminal

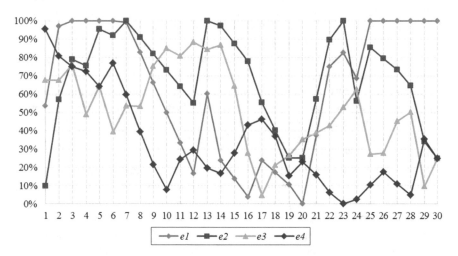

**Fig. 9.22** Case study 2—sequential approach: normalized inventory profiles for utility and product resources

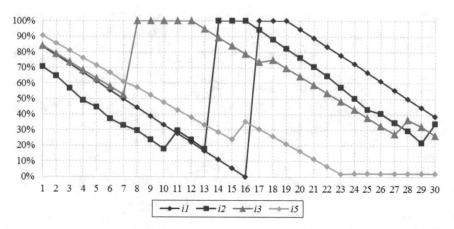

**Fig. 9.23** Case study 2—sequential approach. Performance level profiles for utility units per time period

constraints imposed. However, a much higher inventory level is for utility resource $e1$ is reported, similarly to the solution of the integrated approach. As explained before, this is mainly done for the existence of utility cogeneration units that cogenerate both utilities under different generation ratios (see Table 9.3).

The performance level profiles for active utility units are displayed in Fig. 9.23. It can be seen that the performance level of utility unit $i2$ decreases according to the variation in its operating levels. Utility units $i1$, $i2$ and $i3$ fully recover their performances by undergoing offline cleaning tasks, while utility unit $i5$ undergoes online cleaning in day 16 to partially recover its performance. The performance levels of all operating utility units at the end of the planning horizon remain above 25% (due to the terminal constraints imposed) except for utility unit $i5$ that does not operate in day 30 and therefore terminal constraint was not applied (see Fig. 9.13). In practice, one could perform an offline cleaning on this unit after day 22 to completely restore its performance by the end of the planning horizon.

Figure 9.24 shows the total cost breakdown for the solution of the sequential approach. The operating cost for utility units is 49% which is 5% higher than the percentage of the operating cost of the integrated approach (refer to Fig. 9.10). This is because utility unit $i5$ operates for a longer horizon in sequential approach in comparison with the integrated approach.

Figure 9.25 shows the cost comparison of the solutions derived by following the integrated and the sequential approach. Each cost term for both solutions is divided by the total cost for the sequential approach (which is higher than that of the integrated approach). The major cost difference between the solution of the integrated and the sequential approach is the operating cost for utility units that is about 13%. This difference in the operating cost for the utility system affects strongly the total cost of the solution found by the sequential approach. The extra energy

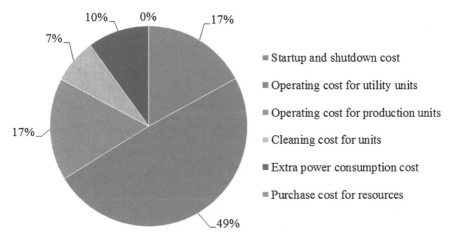

Fig. 9.24 Case study 2—sequential approach: total cost breakdown (percentage)

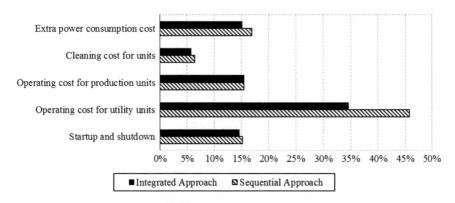

Fig. 9.25 Case study 2: cost term comparison of integrated and sequential approach

consumption cost, cleaning cost and startup and shutdown cost show cost differences of around 1%. The operating cost for production units is almost the same for both approaches.

Figure 9.26 displays the evolution of the total cost value over time for both approaches. This difference significantly increases by the end of the planning horizon. The vertical difference between the two lines in the graph shows the difference of the total cost between the two solutions. In particular, it is observed that the total cost of the solution of the integrated approach is 17% lower than that of the sequential approach demonstrating clearly the benefits of the proposed integrated approach.

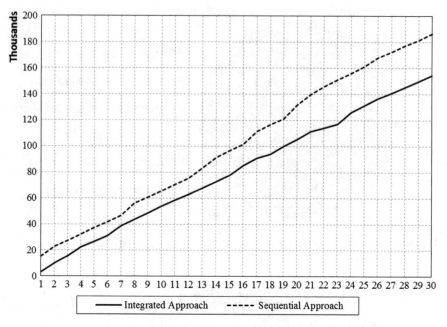

**Fig. 9.26** Case study 2: aggregated total cost over time for integrated and sequential approach

## 9.9 Case Study 3: Integrated Planning of Utility and Production Systems via Rolling Horizon Approach

In this example, the reactive integrated planning problem of utility and production systems through a rolling horizon approach is considered in order to show how the proposed optimization framework can be readily used in a dynamic environment. For the rolling horizon approach, a prediction horizon equal to 15 time periods and a single-period control horizon have been used. A time period is equal to 1 day. The total planning horizon of interest is 30 days, and therefore a total number of 30 iterations have been solved (30 optimization problems). For each iteration, a planning problem for the next 15 time periods is solved with updated information on the current state of the overall system and the demand for product resources. Only the solution of the first time period of the current prediction horizon is applied at each iteration, and the initial state of the next iteration is updated accordingly. In this case study, all utility and production units are subject to alternative condition-based cleaning policies.

## 9.9.1  Description of Case Study 3

This example is a slightly modified version of the previous case study. The main parameters (Table 9.1, 9.2, 9.3, 9.4) and operational costs (Table 9.5) are as before, and the demands for products in the first 30 days is the same as in Case Study 2. In order to apply the rolling horizon approach, there have been considered demands for products for 14 additional time periods (i.e., until day 44) which follow similar a distribution as in the previous periods. Minimum runtime and shutdown times are the same as in the previous examples. Here, all utility and production units are subject to condition-based cleaning, for which there are three alternative cleaning tasks options as before. There is a limited number of available cleaning resources equal to 12 units of cleaning resources.

The parameters that refer to condition-based offline and online cleaning are defined in Table 9.8 are: (i) extra energy consumption limit $\left(v_i^{\max}\right)$; (ii) cumulative time degradation rate $(\delta_i)$; (iii) operating level degradation rate $(\delta_i^q)$; (iv) minimum time between two consecutive online cleaning tasks $\left(\gamma_i^{on}\right)$; (v) recovery factor of the online cleaning $\left(\rho_i^{rec}\right)$; (vii) reduction factor of the operating level for online cleaning $\left(q_i^{on}\right)$; and (vi) resource requirement for online cleaning of a unit $\left(\vartheta_i^{on}\right)$. In addition, the parameters that define the initial state for this case study are given in Table 9.9. Terminal constraints for each prediction horizon are the same as in the previous case study.

## 9.9.2  Results of Case Study 3—Integrated Approach

Figure 9.27 displays how the final plan for the 30-day horizon is constructed through the solution obtained from each iteration (an example of the first three iterations is included). The last Gantt chart in this figure gives the implemented operational and cleaning plan and the total utilization profile of cleaning resources for the planning horizon considered. For the first iteration, the planning problem is

**Table 9.8**  Case study 3: parameters related to the condition-based cleaning of utility and production units

| Parameter | i1 | i2 | i3 | i4 | i5 | i6 | i7 | i8 |
|---|---|---|---|---|---|---|---|---|
| $v_i^{\max}$ | 162 | 153 | 247 | 200 | 210 | 240 | 242 | 247 |
| $\delta_i$ | 9 | 9 | 13 | 10 | 10 | 12 | 11 | 13 |
| $\delta_i^q$ | 6.75 | 6.75 | 9.75 | 7.50 | 7.50 | 9 | 8.25 | 9.75 |
| $\gamma_i^{on}$ | 10 | 10 | 10 | 10 | 10 | 10 | 10 | 10 |
| $\rho_i^{rec}$ | 0.20 | 0.20 | 0.20 | 0.20 | 0.20 | 0.20 | 0.20 | 0.20 |
| $q_i^{on}$ | 0.05 | 0.05 | 0.05 | 0.05 | 0.05 | 0.10 | 0.10 | 0.10 |
| $\vartheta_i^{on}$ | 1 | 1 | 1 | 1 | 1 | 1 | 1 | 1 |

**Table 9.9** Case study 3: initial state of utility and production units

| Parameter | $i1$ | $i2$ | $i3$ | $i4$ | $i5$ | $i6$ | $i7$ | $i8$ |
|---|---|---|---|---|---|---|---|---|
| $\tilde{\rho}_i$ | 9 | 16 | 17 | 4 | 18 | 8 | 5 | 17 |
| $\tilde{\gamma}_i^{on}$ | 22 | 10 | 25 | 41 | 43 | 14 | 39 | 6 |
| $\tilde{\omega}_i$ | 9 | 6 | 17 | 0 | 0 | 8 | 0 | 22 |
| $\tilde{\psi}_i$ | 0 | 0 | 0 | 28 | 9 | 0 | 29 | 0 |
| $\tilde{\rho}_i^{dq}$ | 0 | 0 | 0 | 0 | 0 | 0 | 0 | 0 |
| $\tilde{\beta}_{(e1,z1)}$ | 60 | Units | Initial inventory for utility resource $e1$ | | | | | |
| $\tilde{\beta}_{(e2,z2)}$ | 93 | Units | Initial inventory for utility resource $e2$ | | | | | |
| $\tilde{\beta}_{(e3,z3)}$ | 132 | Units | Initial inventory for product resource $e3$ | | | | | |

solved for time periods 1–15. Only the solution of the first time period is saved. In the second iteration, a new optimization problem for time periods 2–16 is solved having an initial state of the system the past solution for the first time period of the previous iteration. And, the rolling horizon method continues until all 30 iterations are solved (see also Fig. 9.3).

Six offline and seven online cleaning tasks for utility and production units are observed in the implemented Gantt chart. There are some simultaneous condition-based offline cleaning tasks for some units, as listed below: (i) utility unit $i2$ and production unit $i8$ from day 4 and 7; (ii) utility units $i5$ and $i3$ from days 10 and 12; and (iii) utility unit $i1$ and production unit $i6$ in days 19 and 21. In addition, simultaneous online cleanings is observed for utility unit $i1$ and production unit $i6$ in day 7.

Utility unit $i4$, which can only produce utility resource $e1$, operates just in day 1 because utility resource $e1$ has enough supply from the utility units that can cogenerate both utility resources. Utility unit $i5$, which can produce utility resource $e2$, operates for two short-duration period, from day 1 to 5 and from day 15 to 20, because utility units $i2$ and $i1$ are closed for offline cleaning in some of these days. It is also observed that production unit $i7$ remains idle for the whole planning horizon, because the demand for product resources is fully satisfied by the other production units.

The normalized operating level profiles for all units are displayed in Fig. 9.28. In the utility system, utility units $i1$ to $i3$ operate at their maximum operating levels throughout the planning horizon (excluding their cleaning periods). Utility unit $i5$, which can generate only utility resource $e2$, operates in a shorter operating range to satisfy the varied needs for utility resource $e2$. In the production system, production units $i6$ and $i8$ operate at their maximum operating levels almost in all time periods to satisfy the high demand for product resources.

Figure 9.29 depicts the normalized total production profiles for each utility and product resource. The production of each resource is calculated by having the cumulative production of the resource from each unit divided by the maximum total resource production capacity of all units. Similar production trends are observed for

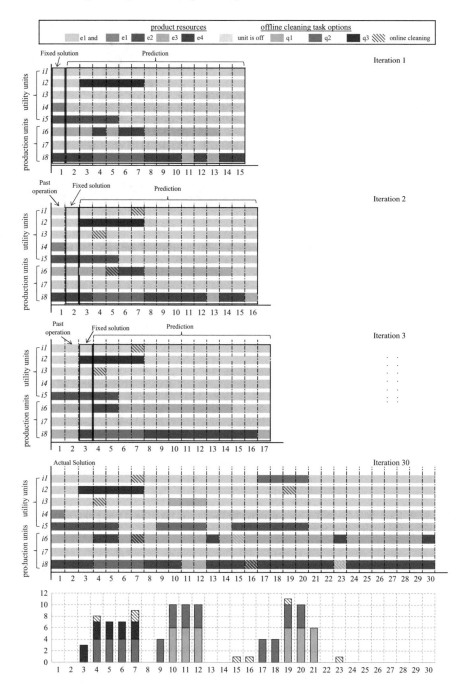

**Fig. 9.27** Case study 3—rolling horizon integrated approach: plan generation via rolling horizon and total utilization profile of cleaning resources

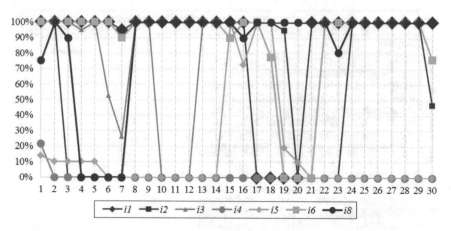

**Fig. 9.28** Case study 3—rolling horizon integrated approach: normalized operating level profiles for utility and production units

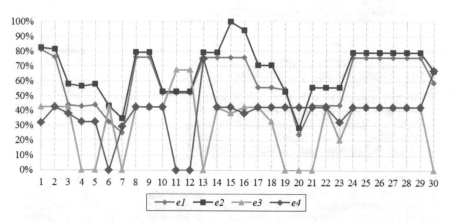

**Fig. 9.29** Case study 3—rolling horizon integrated approach: normalized total production profiles for utility and product resources

utility resources $e1$ and $e2$ mainly due to the presence of three utility units that cogenerate both utility resources. The only differences are observed when utility unit $i5$ operates from day 1 to 5 and from day 15 to 20. There are higher production differences of utility resource $e2$ in comparison to utility resource $e1$. Meanwhile, the production levels for product resources $e3$ and $e4$ from day 8 to 10 and from day 24 to 29 are exactly the same because the upper operating level of utility unit $i6$ that produces product resource $e3$ and the upper operating level of production unit $i8$ that are producing product resource $e4$ in these days are the same (refer to Table 9.1). In addition, when there is no production of product resources in some time periods (e.g., days 4, 5, 7, 13, 19, 20, 21 for product resource $e3$ and days 6,

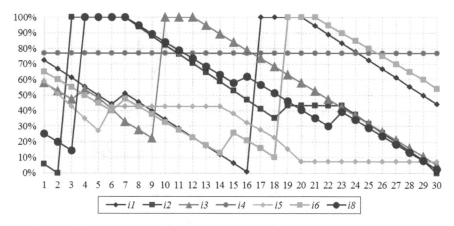

**Fig. 9.30** Case study 3—rolling horizon integrated approach: performance level profiles for utility and production units per time period

11, 12 for product resource $e4$), the demands for product resources are fully satisfied through the inventory tanks for product resources.

The performance level profiles for utility and production units are displayed in Fig. 9.30. It is observed that utility unit $i1$ undergoes online cleaning on day 7 to partially recover its performance and it continues operating until reaching its critical performance level on day 16. The next day, utility unit $i1$ is closed for offline cleaning in order to completely restore its full performance (i.e., clean condition). Production unit $i6$ undergoes two online cleanings (in day 7 and 15) and an offline cleaning on day 19. Utility unit $i5$ shows increased performance degradation from day 14 to 20 due to variation from its reference operating level (refer to Fig. 9.28). It is also observed that utility unit $i5$ reaches a very low performance level and eventually shuts down in day 21. No cleaning task takes place in this unit because it remains idle for the remaining planning horizon. In Fig. 9.31, the performance levels of some operating units in day 30 are below 25% (i.e., terminal constraint) but this is not a violation of the corresponding terminal constraints. The solution of day 30 (including performance level values) has been derived from iteration 30 by solving a planning problem from time period 30 to time period 44, satisfying the terminal constraints for time period 44. In other words, in iteration 30, the terminal constraints apply for the last time period of the planning problem solved (i.e., day 44) and not for the first time period which is day 30.

Figure 9.31 displays the normalized inventory profiles for utility and product resources, having as reference the associated maximum inventory levels. The high inventory levels for utility and product resources in the first period are due to the high initial inventory levels. There are reduced inventory levels for utility resources from day 10–12 and from day 16–18 due to the offline cleaning of some utility units that takes place in these days (see Fig. 9.26). The inventory levels for product resources are reduced on day 4–7 and day 19–21 because of offline cleanings for production units. Recall that all inventory tanks are subject to terminal constraints

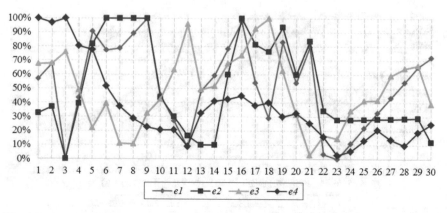

**Fig. 9.31** Case study 3—rolling horizon integrated approach: normalized inventory profiles for utility and product resources

that force the inventory levels in the last time period of each iteration to be 25% of the maximum capacity of the corresponding inventory tank. According to Fig. 9.31, the inventory level for utility resource $e2$ in day 30 is below 25% but this is not a violation of the terminal constraints. The solution of day 30 (including the inventory level values) has been derived from iteration 30 by solving a planning problem from time period 30 to time period 44, satisfying the terminal constraints for the time period 44.

## 9.9.3    Results of Case Study 3—Sequential Approach

Figure 9.32 displays the final Gantt chart and total utilization profile of cleaning resources for the sequential rolling horizon approach. In comparison with the integrated approach, a higher number of offline and online cleaning tasks for utility units is observed. Utility units $i4$ and $i5$ operate in a larger number of time periods than before. Also, production unit $i7$ is utilized in this case, while in the solution from the integrated rolling horizon approach was inactive for the whole planning horizon (see Fig. 9.27). Here, production unit $i7$ operates at the first half of the planning horizon and production unit $i8$ operates mostly in the second half of the planning horizon. This solution also reports a highly increased number of production changeovers in the production units, which in practice can make more complicated the implementation of this plan.

Figure 9.33 displays the aggregated total cost for the integrated and the sequential rolling horizon approach. The total cost of the integrated approach is 14% lower than that of the sequential approach if a zero purchase price is considered, and 32% lower than that of the sequential approach if a purchase price equal to 200 is considered. The results clearly show that the integrated approach can find solutions that are better than those of the sequential approach, even if a zero

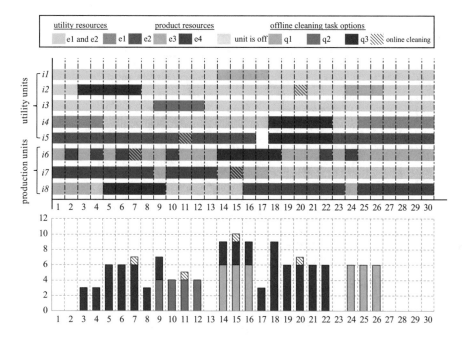

**Fig. 9.32** Case study 3—rolling horizon sequential approach: operational and cleaning plan for production and utility systems and total utilization profile of cleaning resources

purchase price is considered. In practice, penalty or real costs for acquiring utilities from external sources can be very high, since either represent an undesired managerial policy (i.e., dependency on external sources) or high-cost utilities. In this example, the solution following the sequential approach reports a total of 263.8 units of utility resource $e2$ that need to be purchased from external sources, as shown in Table 9.10.

Figure 9.34 shows the cost comparison of the solutions derived by following the integrated and the sequential rolling horizon approach. Note that this figure does not include the purchase cost for resources. As in the previous case study, the highest difference is observed in the operating cost for utility units by about 11%. Extra energy consumption cost difference is at 2%. The cleaning cost and startup and shutdown cost report both a difference of around 0.6%. The operating cost for production units is almost the same for both approaches.

Figure 9.35 shows the CPUs values of each iteration for both approaches. In most of the iterations, the integrated approach shows much higher CPUs values than the sequential approach. The average computational times for the sequential and the integrated approach are 53.9 and 389 CPUs, respectively. It should be clear that the integrated planning problem results in a more complex optimization problem than the sequential planning problem, and therefore higher computational times would be observed for the resolution of the same planning problem. In Fig. 9.35, one can observe that in some iterations, such as iteration 27 and 29, the

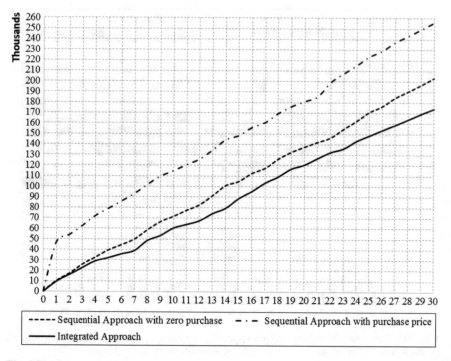

----- Sequential Approach with zero purchase     — · — Sequential Approach with purchase price
——— Integrated Approach

**Fig. 9.33** Case study 3: aggregated total cost for integrated and sequential rolling horizon approaches

**Table 9.10** Case study 3: sequential rolling horizon approach. Utilities purchases

| Utility resource | Amount per time period (in metric units) | | | | | Total (in metric units) |
|---|---|---|---|---|---|---|
| e2 | Day 1 | Day 4 | Day 6 | Day 7 | Day 22 | 263.8 |
|  | 183.6 | 13.9 | 10.4 | 9.2 | 46.8 | |

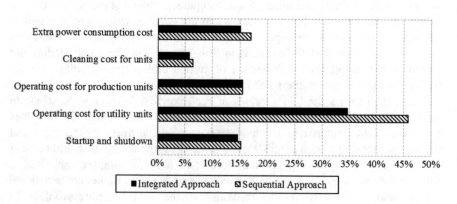

**Fig. 9.34** Case study 3: cost comparison of integrated and sequential rolling horizon approaches

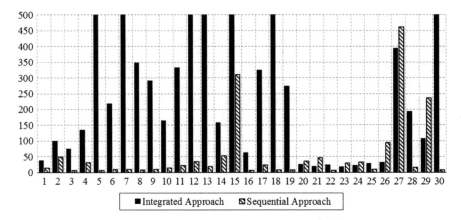

**Fig. 9.35** Case study 3: CPUs values per iteration for integrated and sequential rolling horizon approaches

computational time of the sequential approach is higher than that of the integrated approach. This is due to the fact that the two approaches may not solve exactly the same problem at each iteration (apart from the first iteration), since the planning problem under optimization at each operation depends strongly on the initial state of the system, which in the rolling horizon framework is an optimization output of the previous iteration (apart from the first iteration). Considering the complexity of the integrated planning problems solved in each iteration, the integrated approach reported a very good computational performance.

## 9.10    Concluding Remarks

In this chapter, a rolling horizon optimization framework has been presented for the integrated condition-based planning of utility and production system under uncertainty. Performance degradation and recovery have been considered for both systems. A number of representative case studies showed that the proposed integrated approach can provide significantly better solutions (compared to solutions obtained by sequential approaches) in terms of total costs, and especially in cost terms related to utility units operation, extra energy consumption, and cleaning and startup/shutdown operations. With respect to our previous work, improved unit performance degradation and recovery models that depend on both the cumulative time of operation and the unit operating levels deviation of units have been developed. This is a major step in addressing industrial scenarios. In the case studies solved, we observed that the total cost of the solution of the integrated approach is lower than that of the solution of the sequential approach within a range of 5–32%. This significant reduction in total costs is a direct result of the enhanced energy efficiency of the overall system through the optimized use and consumption of energy (i.e., major parts of the objective function). It has been also demonstrated that unnecessary purchases of resources can be avoided by the proposed integrated approach through the more efficient operation of utility units and the improved utilization handling of energy and material resources. Overall, the proposed approach can result in a cleaner production since energy generation and consumption along with cleaning operations plans (source of waste sources) are optimized. In the longer term, this could result in sustainable production practices. Ongoing research activities focus on the modeling of more complex production processes along with the development of decomposition methods for the effective solution of such highly complicated planning problems.

## 9.11   Nomenclature

### Indices/Sets
$e \in E$   Resources (products and utilities)
$i \in I$   Units (production and utility)
$q \in Q$   Offline cleaning task options
$t \in T$   Time periods
$z \in Z$   Inventory tanks for resources

### Superscripts
es   Earliest
ls   Latest
max   Maximum
min   Minimum
off   Offline
on   Online
s   Startup
f   Shutdown
fix   Fixed
var   Variable
PR   Production system
UT   Utility system
+   Inlet
−   Outlet

### Subsets
$E_i$   Resources that can be produced in unit $i$
$E^{PR}$   Product resources
$E^{UT}$   Utility resources
$I_e$   Units that can produced resource $e$
$I^{SF}$   Units that are subject to startup and shutdown costs

$I^{S-min}$    Units that are subject to minimum runtimes

$I^{F-min}$    Units that are subject to minimum shutdown times

$I^{PR}_e$    Production units that require utility resource $e$ to operate

$Q_i$    Alternative offline cleaning task options for unit $i$

$Z_e$    Inventory tanks that can store resource $e$

$CB_i$    Units $i$ that are subject to condition-based cleaning tasks

$DM_i$    Units $i$ that are under in-progress offline cleaning at the beginning of the planning horizon (information carried over from previous planning horizon)

$FM_i$    Units $i$ that are subject to flexible time-window offline cleaning

$MR_i$    Units $i$ that are subject to maximum runtime constraints

$PR_i$    Production units

$UT_i$    Utility units

## Parameters

$\alpha_{(i,e,e')}$    Coefficient for production unit $i$ that provides the variable needs for utility $e$ for the production of a unit of product $e'$

$\bar{\alpha}_{(i,e,e')}$    Coefficient for production unit $i$ that provides the fixed needs for utility resources $e$ for the production of resources $e'$

$\beta_z^{loss}$    Coefficient of losses in inventory tank $z$

$\gamma_i^{on}$    Minimum time between two consecutive online cleanings in unit $i$

$\delta_i$    Performance degradation rate for unit $i$ due to its cumulative time of operation

$\delta_i^q$    Performance coefficient related to operating level for unit $i$ due to its cumulative deviation from its reference operating level

$\varepsilon_{(e,z,t)}$    Bounds on the total inlet/outlet flow of resource $e$ to/from inventory tank $z$ in time period $t$

$\zeta_{(e,t)}$    Demand for product resource $e \in E^{PR}$ in time period $t$

$\eta_t^{max}$    Limited amount of available resources for cleaning operations in time period $t$

$\vartheta_{(i,q)}^{off}$    Resource requirements for offline cleaning task option $q$ of unit $i$

$\vartheta_i^{on}$    Resource requirements for online cleaning of unit $i$

$\kappa_{(i,t)}$    Bounds on the operating level for utility unit $i \in UT_i$ in time period $t$

$\bar{\kappa}_{(i,e,t)}$    Bounds on the production level of product resource $e \in E^{PR}$ for production unit $i \in PR_i$ in time period $t$

$\lambda_{(e,z)}^B$    Percentage coefficient that determines the minimum level for each resource inventory tank at the end of the prediction horizon (terminal value)

$\lambda_i^U$    Percentage coefficient that determines the maximum extra energy consumption level for operating unit $i$ at the end of the prediction horizon (terminal value)

$M_{(i,t)}, \bar{\mu}_{(i,t)}$    Sufficient big numbers

$\nu_{(i,q)}$    Duration of offline cleaning task option $q$ that could take place in unit $i$

| $\xi_{(e,z)}$ | Bounds on the capacity of inventory tanks $z$ that can store resources $e$ |
|---|---|
| $o_i$ | Maximum runtime for unit $i$ |
| $\pi_i^{on}$ | Percentage modification on the upper operating level of unit $i$ that is under online cleaning |
| $P_{(i,e)}$ | Stoichiometry coefficient that relates the operating level of the utility unit $i$ with the generated amount of each cogenerated utility resource $e$ |
| $\rho_i^{rec}$ | Performance recovery factor of unit $i$ due to online cleaning |
| $\tau_i$ | Time information of cleaning task for unit $i$ |
| $\upsilon_i^{max}$ | Extra energy consumption limit for unit $i$ (performance degradation) |
| $\phi$ | Associated cost coefficients for objective function terms related to utility and production unit $i$ (i.e., variable and fixed operating cost, utilities and products purchase prices, startup and shutdown costs, electricity price, extra energy consumption cost, online and offline cleaning tasks costs) |
| $\psi_i$ | Minimum shutdown idle time for unit $i$ |
| $\omega_i$ | Minimum runtime for unit $i$ |
| $q_{(i,t)}^{ref}$ | Reference operating level for utility unit $i$ per time period |
| $q_{(i,e,t)}^{ref}$ | Reference production level for production unit $i$ that produces product resource per time period |

**Parameters (Initial State of the Overall System)**

| $\tilde{\beta}_{(e,z)}$ | Initial inventory level of resource $e$ in inventory tank $z$ |
|---|---|
| $\tilde{\gamma}_i^{on}$ | Initial state of utility unit $i \in CB_i^{on}$ with respect to its last online cleaning |
| $\tilde{\eta}_{(i,t)}$ | Time periods $t$ for utility unit $i \in DM_i$ that there is a known cleaning resource requirement (in-progress offline cleaning task from previous planning horizon) |
| $\tilde{\rho}_i$ | Initial cumulative time of operation for unit $i$ |
| $\tilde{\rho}_i^q$ | Initial cumulative deviation from the reference operating level for unit $i$ |
| $\tilde{\chi}_i$ | Operating status of unit $i$ just before the beginning of the current planning horizon |
| $\tilde{\psi}_i$ | Total number of time periods at the beginning of the current planning horizon that unit $i$ has been continuously not operating since its last shutdown |
| $\tilde{\omega}_i$ | Total number of time periods at the beginning of the current planning horizon that unit $i$ has been continuously operating since its last startup |

**Continuous Variables (Nonnegative)**

| $B_{(e,z,t)}$ | Inventory level for resource in inventory tank at time period $t$ |
|---|---|
| $B_{(e,z,t)}^-$ | Total outlet flow of resource from inventory tank at time period $t$ |
| $B_{(e,z,t)}^+$ | Total inlet flow of resource to inventory tank at time period $t$ |
| $B_{(e,z,i,t)}^{UT,-}$ | Flow of utility from inventory tank to production unit $i$ at time period $t$ |
| $D_{(i,t)}$ | Cumulative operating level deviation for unit $i$ in time period $t$ |

$\text{NS}^{\text{UT}}_{(e,i,t)}$  Purchases of utility resource to be utilized in production unit $i \in I^{\text{PR}}_e$ in time period $t$

$\text{NS}^{\text{FP}}_{(e,t)}$  Purchases of product resource in time period $t$ (or lost sales)

$\bar{Q}_{(i,t)}$  Operating level of utility unit in time period $t$

$Q_{(i,e,t)}$  Production level of resource from unit $i$ in time period $t$

$R_{(i,t)}$  Cumulative time of operation for unit $i$ in time period $t$

$U_{(i,t)}$  Extra energy consumption (from fully clean condition) of unit $i$ due to its performance degradation

## Binary Variables

$X_{(i,t)}$  = 1, if a unit $i$ is operating during time period $t$

$S_{(i,t)}$  = 1, if a unit $i$ starts up at the beginning of time period $t$

$F_{(i,t)}$  = 1, if a unit $i$ shuts down at the beginning of time period $t$

$V_{(i,t)}$  = 1, if an online cleaning task for unit $i \in \text{CB}^{\text{on}}_i$ occurs in time period $t$

$V^{PR}_{(i,e,t)}$  = 1, if an online cleaning task for production unit $i \in (\text{PR}_i \cap \text{CB}^{\text{on}}_i)$ that produces product resource $e \in E^{\text{PR}}$ takes place in time period $t$

$W_{(i,t)}$  = 1, if an offline cleaning task for unit $i \in (\text{CB}^{\text{off}}_i \cup \text{FM}_i)$ starts at the beginning of time period $t$

$H_{(i,q,t)}$  = 1, if the offline cleaning task option for unit $i \in (\text{CB}^{\text{off}}_i \cup \text{FM}_i)$ starts at the beginning of time period $t$

$Y_{(i,e,t)}$  = 1, if production unit $i \in \text{PR}_i$ produces product resource in time

# References

Kopanos GM, Pistikopoulos EN (2014) Reactive scheduling by a multiparametric programming rolling horizon framework: a case of a network of combined heat and power units. Ind Eng Chem Res 53(11):4366–4386

Kopanos GM, Laínez JM, Puigjaner L (2009) An efficient mixed-integer linear programming scheduling framework for addressing sequence-dependent setup issues in batch plants. Ind Eng Chem Res 48(13):6346–6357

Kopanos GM, Puigjaner L, Georgiadis MC (2010a) Optimal production scheduling and lot-sizing in dairy plants: the yoghurt production line. Ind Eng Chem Res 49(2):701–718

Kopanos GM, Méndez CA, Puigjaner L (2010b) MIP-based decomposition strategies for large-scale scheduling problems in multiproduct multistage batch plants: a benchmark scheduling problem of the pharmaceutical industry. Eur J Oper Res 207(2):644–655

Kopanos GM, Puigjaner L, Maravelias CT (2011a) Production planning and scheduling of parallel continuous processes with product family considerations. Ind Eng Chem Res 50:1369–1378

Kopanos GM, Puigjaner L, Georgiadis MC (2011b) Resource-constrained production planning in semicontinuous food industries. Comput Chem Eng 35(12):2929–2944

Kopanos GM, Puigjaner L, Georgiadis MC (2011c) Production scheduling in multiproduct multistage semicontinuous food processes. Ind Eng Chem Res 50(10):6316–6324

Kopanos GM, Puigjaner L, Georgiadis MC (2012a) Simultaneous production and logistics operations planning in semicontinuous food industries. OMEGA Int J Manage Sci 40:634–650

Kopanos GM, Puigjaner L, Georgiadis MC (2012b) Efficient mathematical frameworks for detailed production scheduling in food processing industries. Comput Chem Eng 42(SI):206–216

Kopanos GM, Xenos DP, Pistikopoulos EN, Thornhill NF (2015) Optimization of a network of compressors in parallel: operational and maintenance planning—the air separation plant case. Appl Energy 146:453–470

Xenos D, Kopanos GM, Cicciotti M, Thornhill NF (2016) Operational optimization of networks of compressors considering condition-based maintenance. Comput Chem Eng 84:117–131

Zulkafli NI, Kopanos GM (2016) Planning of production and utility systems under unit performance degradation and alternative resource-constrained cleaning policies. Appl Energy 183:577–602

Zulkafli NI, Kopanos GM (2017) Integrated condition-based planning of production and utility systems under uncertainty. J Clean Prod 167:776–805

# Part VI
# Conclusions and Outlook

# Chapter 10
# Conclusions and Outlook

## 10.1 Conclusions

The main aim of this book has been to establish mathematical programming techniques and solution approaches for the efficient solution of complex production scheduling and planning problems. For this reason, a number of real-life case studies, contemplating representative sectors and significant process industries (chemical, pharmaceuticals, food and beverage industries), have been addressed and solved by new mathematical programming frameworks.

Part I identifies the major challenges to be addressed through an extensive state-of-the-art review (Chap. 2). Although production planning and scheduling has become the subject of intensive research, an attentive review reveals the areas where new contributions are needed for a major impact in real-world applications. In Chap. 3, the fundamental theory and concepts behind the optimization methods and tools used throughout the book is briefly described.

Part II deals with continuous production processes. More specifically, in Chap. 4, an industrial case study considering the simultaneous planning and scheduling problem in the bottling stage of a real-life beer industry, producing hundreds of final products, has been addressed. A special feature of the problem in question is that final products can be classified into product families. The grouping into families is based on various criteria, including product similarities, processing similarities, and/or changeover considerations. A hybrid discrete/continuous-time mathematical approach to the simultaneous production planning and scheduling of continuous parallel units producing a large number of final products that can be classified into product families has been developed. In contrast with previous research works, a more general case has been considered based on: (i) product families, (ii) short planning periods that may lead to idle units for entire periods, (iii) changeovers spanning multiple periods, and (iv) maintenance activities. The proposed approach also addresses appropriately aspects such as changeover carryover and crossover, thereby leading to solutions with higher utilization of resources. Very good solutions

© Springer Nature Switzerland AG 2019
G. M. Kopanos and L. Puigjaner, *Solving Large-Scale Production Scheduling and Planning in the Process Industries*,
https://doi.org/10.1007/978-3-030-01183-3_10

to problems with hundreds of products can be obtained within 5 CPU min, while optimal solutions can also be found in a reasonable time. Furthermore, the proposed formulation yields solutions which are substantially better than the ones obtained using commercial tools, suggesting that MIP methods can be used to address large-scale problems of practical interest.

Part III deals with food process industries that combine batch and continuous operation modes in their overall production process. In Chap. 5, a multiproduct, multistage, semicontinuous ice-cream production facility is considered. A new MIP framework and a solution strategy have been presented for the optimal production scheduling of this production facility. Although the proposed mathematical formulation is well suited to the ice-cream production facility considered, it could be also used, with minor modifications, in scheduling problems arising in other semicontinuous industries with similar processing features. The overall mathematical framework relies on an efficient modeling approach of the sequencing decisions, the integrated modeling of all production stages and the inclusion of strong valid integer cuts in the MIP formulation. The simultaneous optimization of all processing stages increases the plant production capacity, reduces the production cost for final products, and facilitates the interaction among the different departments of the production facility. The proposed MIP formulation and the anticipated solution methodology results in very low computational times for the several problem instances solved.

In Chap. 6, a multiproduct semicontinuous yogurt production facility, where labor (i.e., the number of available workers) is a limited resource, is studied. Production planning in semicontinuous processing plants typically deals with a large number of products, however, many products appear with similar characteristics, and therefore final products can be grouped into product families. Thus, the production planning problem under question could be partially focused on product families rather than on each product separately, following a similar modeling concept to Chap. 4. A general MIP approach has been presented for the resulting resource-constrained production. Quantitative as well as qualitative optimization goals are included in the proposed model. The definition of product families significantly reduces the size of the underlying mathematical model and, thus, the necessary computational effort without sacrificing any feasibility constraints. A number of cases studies, also considering unexpected event scenarios (i.e., workers absence, and products orders modifications), have been solved in reasonable computational time.

Chapter 7 addresses the production and logistics operations planning in large-scale single- or multisite semicontinuous process industries. A novel mixed discrete continuous-time mixed integer programming model, based on the concept of families of products, for the problem in question has been developed. A remarkable feature of the proposed approach is that in the production planning problem timing and sequencing decisions are taken for product families rather than for products. However, material balances are realized for every specific product, thus permitting the detailed optimization of production, inventory, and transportation costs. Additionally, alternative transportation modes are considered for the delivery of final products from production sites to distribution centers, a reality that most of the current approaches totally neglect. The efficiency and the applicability of the proposed approach are demonstrated by solving to optimality two industrial-size case studies, for an emerging real-life dairy industry. It is worth noting that despite the complexity of the problems addressed, the proposed approach appears a remarkable computational performance.

Part IV deals with scheduling in batch processes. In Chap. 8, a real-life multiproduct multistage pharmaceuticals scheduling problem is considered. A systematic two-stage iterative solution strategy, based on mathematical programming, has been developed. More specifically, the proposed solution strategy consists of a constructive step, wherein a feasible and initial solution is rapidly generated by following an iterative insertion procedure, and an improvement step, wherein the initial solution is systematically enhanced by implementing iteratively several rescheduling techniques; based on the mathematical model. A salient feature of the proposed approach is that the scheduler can maintain the number of decisions at a reasonable level thus reducing appropriately the search space. This usually results in manageable model sizes that often guarantees a more stable and predictable optimization model behavior. Several challenging large-scale scheduling problem instances, considering alternative optimization goals, of a pharmaceuticals production facility have been solved. Also, it is worth mentioning that a new precedence concept (i.e., the unit-specific general precedence that is included in Appendix B) has been developed in order to cope with objectives containing changeover issues.

Finally, Part V deals with the simultaneous operational planning and maintenance of the utility and production systems of a process industry. In Chap. 9, a general optimization framework is presented with the main purpose of reducing the energy needs and material resources utilization of the overall system. The proposed mathematical model focuses mainly on the utility system and considers for the utility units: (i) unit commitment constraints, (ii) performance degradation and recovery, (iii) different types of cleaning tasks (online or offline, and fixed or flexible time window), (iv) alternative options for cleaning tasks in terms of associated durations, cleaning resources requirements and costs, and (v) constrained availability of resources for cleaning operations. The optimization function includes the operating costs for utility and production systems, cleaning costs for utility systems, and energy consumption costs. Several industrial-inspired case studies are presented in order to highlight the applicability and the significant benefits of the proposed approach. In particular, in comparison with the traditional sequential planning approach for production and utility systems, the proposed integrated approach can achieve considerable reductions in startup/shutdown and cleaning costs, and most importantly in utilities purchases, as it is shown in one of the case studies.

## 10.2   Outlook

This book focused on the development and presentation of effective modeling concepts and mathematical programming approaches to efficiently tackle real-life industrial planning and/or scheduling problems in the process industries, in an attempt to make more attractive the mathematical approaches and bridge the gap between planning and scheduling theory and practice. A range of issues requiring further investigation has been revealed in the course of this work. In particular:

- Further study and improvement of the mathematical-based solution techniques. For instance, the MIP-based solution strategy, presented in Chap. 8, could be combined with metaheuristics in an attempt to reduce the computational burden of large-scale optimization problems.
- Since process industries are dynamic in nature, the consideration of the uncertainty also arises as a challenging research task. As briefly described in Sect. 2.4, proactive (e.g., stochastic programming or parametric optimization) or reactive (e.g., full or partial replanning) approaches could be used (Acevedo and Pistikopoulos 1997a, b; Pistikopoulos et al. 2002; Balasubramanian and Grossmann 2004; Dua et al. 2008).
- Continual improvement in mathematical problem formulation and preprocessing to improve relaxation characteristics, and tailor-made solution procedures for problems with relatively large integrality gaps.
- More attention should be given to the multisite production problem. A major task should be the simultaneous optimization of production and logistics operations across multiple production facilities and distribution centers, in order to enhance the overall performance, responsiveness, and profitability of the enterprise.
- Regarding multisite problems, when an extremely large number of final products, production sites, and distribution centers are present, appropriate modeling frameworks and solution strategies should be devised in order to tackle efficiently these highly complicated optimization problems.
- Implementation of the proposed models, after further improvements, in a computer-aided advanced scheduling and planning system.
- There are process industries that have received little attention; regarding scheduling and/or planning research, such as the food industries considered in this thesis. For instance, scheduling and planning approaches in the ceramics and tiles process industry are rather poor, despite the fact that there are many optimization challenges.
- New efficient approaches to integrate scheduling decisions into SC design may be further explored.
- More rigorous treatment of the financial aspects as scheduling and planning become integrated (Badell et al. 2004; Guillén et al. 2007; Laínez et al. 2007). And, further extension of the proposed mathematical approaches to address

environmental and sustainability considerations (Stefanis et al. 1997), thus necessitating the development of multiobjective optimization frameworks (Puigjaner et al. 2009).

- Solution techniques and concepts developed for the planning and scheduling of the process industries can be implemented in other classes of planning or scheduling problems. For instance, process scheduling concepts could be applied into project scheduling problems or into typical manufacturing industries.

The implementation of production scheduling and planning software is a highly challenging task that require the close involvement of the industrial users. Mathematical models that may seem perfect and efficient in paper may not find application in the real scenario for several reasons. For example, some data required by the models may be hard to obtain or difficult to get good estimate values, and frequently the industrial practice requires the consideration of multiple qualitative rules or aspects that may be challenging to be quantified and integrated into the overall optimization framework in an efficient manner. The development of production scheduling and planning tools is a highly challenging activity that requires human resources with a wide variety of knowledge and skills related to data science, optimization, operational research, and computers science, etc. Last but not least, an extremely challenging task is to deliver such tools that would get the acceptance from the industrial sector. In order to achieve this it is essential to have close contact with the industrial experts and final users and provide them with appropriate training of how to use the tool, since in most of the cases although the end users have excellent experience in the production process, they have very limited knowledge of the principles of production scheduling and planning.

# References

Acevedo J, Pistikopoulos EN (1997a) A hybrid parametric/stochastic programming approach for mixed-integer linear problems under uncertainty. Ind Eng Chem Res 36(6):2262–2270

Acevedo J, Pistikopoulos EN (1997b) A multiparametric programming approach for linear process engineering problems under uncertainty. Ind Eng Chem Res 36(3):717–728

Badell M, Romero J et al (2004) Planning, scheduling and budgeting value-added chains. Comput Chem Eng 28(1–2):45–61

Balasubramanian J, Grossmann IE (2004) Approximation to multistage stochastic optimization in multiperiod batch plant scheduling under demand uncertainty. Ind Eng Chem Res 43 (14):3695–3713

Dua P, Kouramas K, Dua V, Pistikopoulos EN (2008) MPC on a chip—recent advances on the application of multi-parametric model-based control. Comput Chem Eng 32(4–5):754–765

Guillén G, Badell M, Puigjaner L (2007) A holistic framework for short-term supply chain management integrating production and corporate financial planning. Int J Prod Econ 106:288–306

Laínez JM, Guillén-Gonsálbez et al (2007) Enhancing corporate value in the optimal design of chemical supply chains. Ind Eng Chem Res 46(23):7739–7757

Pistikopoulos EN, Dua V et al (2002) On-line optimization via off-line parametric optimization
  tools. Computers and Chemical Engineering 26(2):175–185
Puigjaner L, Laínez JM, Álvarez CR (2009) Tracking the dynamics of the supply chain for
  enhanced production sustainability. Ind Eng Chem Res 48(21):9556–9570
Stefanis SK, Livingston AG, Pistikopoulos EN (1997) Environmental impact considerations in the
  optimal design and scheduling of batch processes. Comput Chem Eng 21(10):1073–1094

# Appendix A
# The Unit-Specific General Precedence Concept

**Abstract** This appendix contains a detailed description of the unit-specific general precedence concept used in the MIP-based solution strategy presented previously in Chap. 8.

## A.1 Introduction

Continuous-time representation strategies based on the precedence relationships between batches to be processed have been developed to deal with the process scheduling problem. In these mathematical formulations, model variables and constraints enforcing the sequential use of shared resources are explicitly employed, and, therefore, sequence-dependent setups (changeovers) can be treated straightforwardly Méndez et al. (2006). The three different precedence-based approaches that can be found in the literature are: (i) the immediate precedence, (ii) the unit-specific immediate precedence, and (iii) the general precedence. Immediate (or local) Precedence (IP) explores the relation between each pair of consecutive orders in the production schedule time horizon without taking into account whether the orders are assigned to the same unit. Unit-Specific Immediate Precedence (USIP) is based on the immediate precedence concept. The difference is that it only takes into account the immediate precedence of the orders that are assigned to the same processing unit. General (or global) Precedence (GP) generalizes the precedence concept by exploring the precedence relations of each batch, taking into account all the remaining batches and not only the immediate predecessor Méndez and Cerdá (2003a). The last approach results in a lower number of binary variables and, compared with the other two approaches, it significantly reduces the computational effort on average. However, it cannot cope with changeovers issues explicitly (especially if there are changeover

© Springer Nature Switzerland AG 2019                                                        253
G. M. Kopanos and L. Puigjaner, *Solving Large-Scale Production
Scheduling and Planning in the Process Industries*,
https://doi.org/10.1007/978-3-030-01183-3

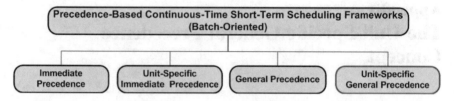

**Fig. A.1** Current and proposed precedence-based scheduling frameworks

times greater than a batch processing time), as it is clearly demonstrated in the illustrative example. Moreover, scheduling models based on the GP notion cannot be used to address problems with sequence-dependent changeover costs because the global sequencing variables are active for all the batch pairs assigned to the same unit. In order to address this limitation, a new precedence-based scheduling formulation, the Unit-Specific General Precedence (USGP), is developed here. Figure A.1 shows the precedence-based frameworks for the scheduling of batch processes.

## A.2    Problem Statement

The scheduling problem in single-stage multiproduct batch plants with different processing units working in parallel is addressed here. Batch to the unit assignment and batch sequencing in order to meet a production goal constitutes the understudy scheduling problem. Changeover times, which greatly increase the complexity of the problem, are explicitly considered. The main assumptions of the proposed model include the following:

(i)   Only single-stage product orders are considered.
(ii)  An equipment unit cannot process more than one batch at a time.
(iii) To begin another task in a processing unit, the current task must have been completed (i.e., non-preemptive operation mode).
(iv)  Processing times, unit setup times, and changeover times and/or costs and due dates are deterministic.
(v)   Unforeseen events, such as unit breakdowns, that may disrupt the normal plant operation do not appear during the scheduling time horizon.
(vi)  No resource constraints except for equipment availability are taken into account.
(vii) Product batch sizing is carried out beforehand, and thus batch sizes are known a priori.

In industrial batch plants, assumption (iii) is frequently satisfied. This is not the case, however, for assumption (i), since there are many industrial applications that are multistage. The proposed model can be appropriately modified to deal with multistage plants (see Chap. 8). By adding a set of resource constraints similar to the ones that were reported in the work by Marchetti and Cerdá (2009b), assumption (vi) can be satisfied. If assumptions (iv) and/or (v) are relaxed, then uncertainty should be included in the optimization procedure. There are generally two ways of treating unexpected events: proactively (offline) or reactively (online). The proposed scheduling framework can be easily adapted to both cases. Finally, assumption (vii) is in accordance with the sequential modeling strategy in scheduling problems, in which first the lot-sizing problem is solved and then, once the number and sizes of batches are known, the pure scheduling problem is solved. This approach probably results in less optimal solutions than the monolithic approach, in which lot-sizing and scheduling are simultaneously optimized. However, the sequential scheduling approach is less computationally expensive, and in some cases, it can be viewed as a good approximation to the industrial reality.

## A.3  The Unit-Specific General Precedence Scheduling Model

In this section, the proposed MIP scheduling model is described in detail. The concept of the USGP is also introduced and explained. In the proposed mathematical formulation, constraints have been grouped according to the type of decision (e.g., assignment, timing, and sequencing) they are imposed on.

**Allocation Constraints**  Constraint set (A.1) presents the unit allocation constraints for every order $i$. As this expression states, each order $i$ can be assigned to only one processing unit $j$ or to none (permitting unsatisfied demand). $Y_{ij}$ represents the binary decision of whether to assign a product order $i$ to a processing unit $j$ or not. $Y_{ij}$ is active, i.e. $Y_{ij} = 1$, whenever product $i$ is allocated to unit $j$, otherwise, it is set to zero. Let $J_i$ denote the set of units $j$ that can process product $i$. By changing the inequality to an equality total demand satisfaction is imposed.

$$\sum_{j \in JI_i} Y_{ij} \leq 1 \quad \forall i \qquad (A.1)$$

**Timing Constraints**  The completion time $C_i$ for batch $i$, when it is assigned to unit $j$, should be greater than the summation of its corresponding processing time $\tau_{ij}$ and setup time $\tau_{ij}$ in this unit $j$. The maximum of the ready unit time $\tau_{ij}$ and the release order time $o_i$ are also added to this summation as the following equation states.

$$C_i \geq \sum_{j \in J_i} (\max[\in_j, o_i] + \tau_{ij} + \pi_{ij}) Y_{ij} \quad \forall i \qquad (A.2)$$

**Sequencing-Timing Constraints**  Constraint set (A.3) expresses the order sequencing constraints between two orders, $i$ and $i'$. This equation is formulated as a big-M constraint, and it is activated for every task $i'$ that is processed after task $i$, when both are assigned to the same unit $j$ (i.e., $X_{ii'j} = 1$).

$$C_i + \gamma_{ii'j}\bar{X}_{ii'j} \le C_{i'} - \tau_{i'j} - \pi_{i'j} + M(1 - X_{ii'j}) \quad \forall i, i' \ne i, j \in (J_i \cap J_{i'}) \quad (A.3)$$

Binary variable $\bar{X}_{ii'j}$ defines the immediate precedence of two tasks $i$ and $i'$, when both are assigned to the same unit $j$. If two orders $i$ and $i'$ are allocated to the same processing unit $j$ and order $i'$ is processed directly after order $i$, then $\bar{X}_{ii'j} = 1$. Hence, this formulation allows us to consider explicitly and efficiently the changeover times and/or costs.

**Sequencing-Allocation Constraints** In order to assess binary variable $\bar{X}_{ii'j}$ with the sequencing-assignment binary variable $X_{ii'j}$, a set of additional constraints, presented below, is needed. As mentioned above, the binary variable $X_{ii'j}$ only stands for two products, $i$ and $i'$, that are assigned to the same equipment unit $j$. In disjunctive programming this statement can be expressed as follows:

$$X_{ii'j} \Rightarrow [Y_{ij} \wedge Y_{i'j}] \quad \forall i, i' \ne i, j \in (I_i \cap J_{i'})$$

Later, the aforementioned disjunctive programming expression is decomposed into constraints (A.4) and (A.5). It can be clearly seen that $X_{ii'j}$ may take the value of 1 only if both orders $i$ and $i'$ are into the same unit $j$; otherwise, it is set to zero without exploring the sequencing of the orders further.

$$Y_{ij} + Y_{i'j} \le 1 + X_{ii'j} + X_{i'ij} \quad \forall i, i', j \in (J_i \cap J_{i'}) : i' > i \quad (A.4)$$

$$2(X_{ii'j} + X_{i'ij}) \le Y_{ij} + Y_{i'j} \quad \forall i, i', j \in (J_i \cap J_{i'}) : i' > i \quad (A.5)$$

In order to explicitly tackle scheduling problems with changeover times and/or costs, the immediate precedence of every pair of orders must be assessed. We now describe our approach. Obviously, two orders $i$ and $i'$ may be consecutive only in the case that the sequencing-allocation binary variable is $X_{ii'j} = 1$ and when there is no other order $i''$ between orders $i$ and $i'$, and vice versa. In disjunctive programming, this expression can be stated as follows:

$$X_{ii'j} \wedge \neg \left[ \bigvee_{i'' \ne [i,i']} (X_{ii'i''j} - X_{i'i''j}) \right] \Leftrightarrow \bar{X}_{ii'j} \quad \forall i, i' \ne i, j \in (J_i \cap J_{i'})$$

Figure 8.1, found on page 164, shows an illustrative example of the concept of this expression. It can be seen that if the total number of batches that follow batch $i$, excluding batch $i'$ is equal to the total number of batches that follow batch $i'$, excluding batch $i$, then batches $i$ and $i'$ are consecutive. For the sake of simplicity, only one processing unit is considered. As a unique machine case is studied, the unit index $j$ is omitted in this example. Constraints (A.6) and (A.7) correspond to the mathematical formulation of the aforementioned disjunctive programming expression. Constraint set (A.6) states that the auxiliary variable $z_{ii'j}$ will be set to zero only if batch $i$ is processed before batch $i'$ and they are allocated to the same equipment unit $j$ (i.e., $X_{ii'j} = 1$) and simultaneously $\sum_{i'' \neq [i,i']} X_{ii''j} - X_{i'i''j} = 0$ (see Fig. 8.1).

$$Z_{ii'j} = \sum_{i'' \neq [i,i']} (X_{ii''j} - X_{i'i''j}) + M(1 - X_{ii'j}) \quad \forall i, \ i' \neq i, \ j \in (J_i \cap J_{i'}) \tag{A.6}$$

If the position difference variable $Z_{ii'j}$ is equal to zero, i.e., when order $i$ has been processed exactly before order $i'$, constraint set (A.7) activates the binary variable $Seq_{ii'j}$, i.e., $\bar{X}_{ii'j} = 1$. Therefore, the consecutiveness of the orders is assessed and sequence-dependent setup time and/or cost issues can be effectively treated.

$$Z_{ii'j} + \bar{X}_{ii'j} \geq 1 \quad \forall i, \ i' \neq i, \ j \in (J_i \cap J_{i'}) \tag{A.7}$$

**Objective Function** Different objective functions can be optimized by using the proposed scheduling framework. For instance, the earliness and the tardiness for every product order $i$ are given by

$$E_i \geq \delta_i - C_i \quad \forall i \tag{A.8}$$

$$T_i \geq C_i - \delta_i \quad \forall i \tag{A.9}$$

The minimization of a combined function of earliness and tardiness, as given in equation (A.10), is one of the most widely used objective functions in the scheduling literature. It is also known as weighted lateness. The weighting coefficients $\alpha_i$ and $\beta_i$ are used to specify the significance of order earliness or tardiness respectively.

$$\min \sum_i (\alpha_i E_i + \beta_i T_i) \tag{A.10}$$

If tardiness is not permitted $(T_i = 0)$ for any order $i$, the aforementioned objective function can be substituted by the maximization of the order completion time, $Ci$:

$$\max \sum_i C_i \tag{A.11}$$

This objective function is identical to the minimization of earliness. Note that if changeover costs are proportional to the changeover times, then the minimization of earliness will correspond to minimization of changeover costs. Alternative objective functions can be also used (e.g., minimization of makespan, total costs minimization, changeover costs minimization, maximization of profit).

At this point, it should be noted that general precedence-based formulations cannot optimize changeover costs, because batch consecutiveness is not assessed explicitly. In the process industries, changeover considerations and optimization are of great importance. Moreover, since scheduling constitutes a part of the supply chain network optimization problem which ought to meet financial goals, it must also be examined and optimized considering financial and economic issues. In view of this industrial strategy in the contemporary highly competitive market environment, the significant advantages of adopting the proposed scheduling framework (to explicitly deal with changeover issues) are clear. What is more, general precedence models may generate suboptimal solution when changeover times are greater than processing times, as will be demonstrated in the illustrative example presented.

## A.4   A General Precedence Scheduling Model

In this section, the representative precedence-based mathematical formulation of Méndez and Cerdá (2003a) is presented.

$$\sum_{j \in J_i} Y_{ij} = 1 \quad \forall i \tag{A.12}$$

$$C_i \geq \sum_{j \in J_i} (\max[\varepsilon_j, o_i] + \tau_{ij} + \pi_{ij}) Y_{ij} \quad \forall i \tag{A.13}$$

$$C_i + \gamma_{ii'j} \leq C_{i'} - \tau_{i'j} - \pi_{i'j} + M(1 - X_{ii'}^{GP}) M(2 - Y_{ij} - Y_{i'j}) \\ \forall i, i', j \in (J_i \cap J_{i'}) : i' > i \tag{A.14}$$

$$C_{i'} + \gamma_{i'ij} \leq C_i - \tau_{ij} - \pi_{ij} + MX_{ii'}^{GP} + M(2 - Y_{ij} - Y_{i'j}) \\ \forall i, i', j \in (J_i \cap J_{i'}) : i' > i \tag{A.15}$$

Constraint set (A.12) corresponds to the unit allocation constraints, and constraint set (A.13) defines the completion time for every product. Constraint sets (A.14) and (A.15) define the relative sequencing of product batches at each processing unit. These sets of big-M constraints enforce the starting time of a product $i'$ to be greater than the completion time of whichever product $i$ processed beforehand. Note that $X_{ii'}^{GP}$ corresponds to the global sequencing binary variable, which is active (i.e., $X_{ii'}^{GP} = 1$) for all products $i'$ that are processed after product $i$.

## A.5   Illustrative Example

A modified version of a plastic compounding plant, first introduced by Pinto and Grossmann (1995), is used as a simple illustrative example. Table A.1 shows the data for this scheduling problem. The optimization goal is to minimize earliness and tardiness (where $\alpha_i = 1$ and $\beta_i = 5$).

**Table A.1** Data for the motivating example

| Order | Due date (day) | Processing times (days) | | | |
|---|---|---|---|---|---|
| | | Unit 1 | Unit 2 | Unit 3 | Unit 4 |
| 1 | 15 | 1.538 | | | 1.194 |
| 2 | 30 | 1.500 | | | 0.789 |
| 3 | 22 | 1.607 | | | 0.818 |
| 4 | 25 | | | 1.564 | 2.143 |
| 5 | 20 | | | 0.736 | 1.017 |
| 6 | 30 | 5.263 | | | 3.200 |
| 7 | 21 | 4.865 | | 3.025 | 3.214 |
| 8 | 26 | | | 1.500 | 1.440 |
| 9 | 30 | | | 1.869 | 2.459 |
| 10 | 29 | | 1.282 | | |
| 11 | 30 | | 3.750 | | 3.000 |
| 12 | 21 | | 6.796 | 7.000 | 5.600 |
| 13 | 30 | 11.250 | | | 6.716 |
| 14 | 25 | 2.632 | | | 1.527 |
| 15 | 24 | 5.000 | | | 2.985 |
| Setup times | (days) | 0.180 | 0.175 | | 0.237 |

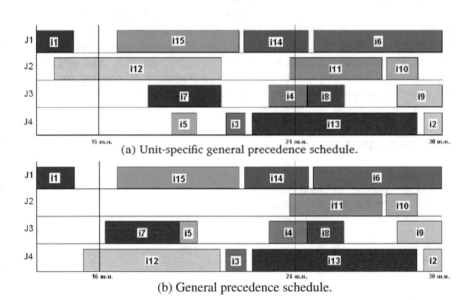

(a) Unit-specific general precedence schedule.

(b) General precedence schedule.

**Fig. A.2** Motivating example schedules: general precedence fault demonstration

First, the addressed scheduling problem is solved without changeover times considerations (i.e., $i'j = 0$). The objective function value is equal to 6.681 m.u.; the schedule obtained is shown in Fig. A.2a. Afterward, the scheduling problem is solved by considering changeover times only between order $i_{12}$ and order $i_{10}$ ($i_{12}$ $i_{10}$ $j = 9$). As the schedule in Fig. A.2a shows, order $i_{12}$ and order $i_{10}$ are not consecutive; thus, someone will expect to obtain the same schedule even if a changeover time is assigned between these orders. However, this is not the case when the general precedence model is applied. Figure A.2b illustrates the schedule obtained, which is, perhaps unsurprisingly, different from the previous one. Its objective function value is equal to 8.472 m.u. (26% worse).

The cause of this fault is that the general precedence sequencing constraints take into account all the sequence-dependent changeover times (and not only between consecutive orders) of the orders assigned to the same unit $j$. This point can be clearly seen by observing constraints (A.14) and (A.15). It can be clearly seen that changeover times, $\gamma_{ii'j}$, are taken into account whenever the general precedence sequencing variable $X_{ii'}^{GP}$ is active (i.e., $X_{ii'}^{GP} = 1$).

Therefore, coming back to the illustrative example, because $X_{i_{12}i_{10}} = 1$ (in the schedule in Fig. A.2a order $i_{12}$ is processed before order $i_{10}$, although they are not consecutive orders) the sequence-dependent changeover time $\gamma_{i_{12}i_{10}} = 9$ (of two no-consecutive orders) is incorrectly taken into account. In other words, the general precedence model will try to make an order completion time plus the changeover time less than or equal to the starting time of any other following order, and consecutiveness is not explicitly considered.

All in all, in cases, that there exist some sequence-dependent changeover times higher than some batch processing times, as in this illustrative example, general precedence may result in a suboptimal solution. If all changeover times are lower than all the orders processing times then general precedence is valid and can be implemented. Nevertheless, note that sequence-dependent changeover cost issues still cannot be addressed explicitly by general precedence models.

## A.6   Nomenclature

**Indices/Sets**
$i, i', i'' \in I$    product orders (products)
$j \in J$              processing units (units)

**Subsets**
$J_i$    available units $j$ to process product $i$

**Parameters**
$\alpha_i$      weighing coefficient for earliness for product $i$
$\beta_i$      weighing coefficient for tardiness for product $i$
$\gamma_{ii'j}$    sequence-dependent setup (changeover) time between products $i$ and $i'$ in unit $j$
$\delta_i$      due date for product $i$
$\varepsilon_j$      time point that unit $j$ is available to start processing

$M$      a big number
$o_i$      release time for product $i$
$\pi_{ij}$      sequence-independent setup time of product $i$ in unit $j$
$\tau_{ij}$      processing time for product $i$ in unit $j$

**Continuous Variables**
$C_i$      completion time of product $i$
$E_i$      earliness for product $i$
$T_i$      tardiness for product $i$
$Z_{ii'j}$      allocation position difference between products $i$ and $i'$ in unit $j$

**Binary Variables**
$X_{ii'j} = 1$      for every product $i$ that is processed before product $i'$ in unit $j$
$\bar{X}_{ii'j} = 1$      1 if product $i$ is processed exactly before product $i'$ in unit $j$
$X_{ii'}^{GP} = 1$      for every product $i$ that is processed before $i'$
$Y_{ij} = 1$      if product $i$ is assigned to unit $j$

# References

Marchetti PA, Cerdá J (2009a) A continuous-time tightened formulation for single-stage batch scheduling with sequence-dependent changeovers. Ind Eng Chem Res 48(1):483–498.
Méndez CA, Cerdá J (2003a) An MILP continuous-time framework for short-term scheduling of multipurpose batch processes under different operation strategies. Optim Eng 4(1–2):7–22.
Méndez CA, Cerdá J (2003b) Dynamic scheduling in multiproduct batch plants. Comput Chem Eng 27:1247–1259.

# Appendix B
# Data for the Resource-Constrained Yogurt Production Process

***Abstract*** This appendix contains the main processing data for the yogurt production process described and studied in Chap. 6. More specifically, Table B.1 provides the main data for all final yogurt products including: (i) the assigned products to families set $P_f$, (ii) the product cup weight, (iii) the product inventory cost $\xi_{pn}$, and (iv) the minimum production amount for any product. Packing rates for everyproduct can be found in Table B.2. Changeover times among families are given in Table B.3. Changeover costs are related to changeover times, according to Table B.4, but setups are irrelevant. Finally, Table B.5 provides the production targets (for Case Study I and II).

© Springer Nature Switzerland AG 2019
G. M. Kopanos and L. Puigjaner, *Solving Large-Scale Production Scheduling and Planning in the Process Industries*,
https://doi.org/10.1007/978-3-030-01183-3

**Table B.1** Main data for final products

| Product | Family | Weight (kg) | Inv. cost (€) | Min. run (kg) | Product | Family | Weight (kg) | Inv. cost (€) | Min. run (kg) | Product | Family | Weight (kg) | Inv. cost (€) | Min. run (kg) |
|---|---|---|---|---|---|---|---|---|---|---|---|---|---|---|
| P01 | F01 | 0.600 | 9.00 | 82.80 | P32 | F08 | 0.600 | 1.65 | 82.80 | P63 | F17 | 0.750 | 0.90 | 27.00 |
| P02 | F01 | 0.600 | 7.50 | 82.80 | P33 | F08 | 0.600 | 1.65 | 82.80 | P64 | F17 | 0.750 | 0.60 | 27.00 |
| P03 | F01 | 0.600 | 6.75 | 82.80 | P34 | F09 | 0.450 | 0.45 | 62.10 | P65 | F17 | 0.750 | 0.90 | 27.00 |
| P04 | F01 | 0.600 | 6.00 | 82.80 | P35 | F09 | 0.450 | 0.45 | 62.10 | P66 | F17 | 0.750 | 0.60 | 27.00 |
| P05 | F01 | 0.600 | 6.00 | 82.80 | P36 | F09 | 0.150 | 1.20 | 20.70 | P67 | F17 | 0.750 | 0.90 | 27.00 |
| P06 | F02 | 0.600 | 6.00 | 82.80 | P37 | F09 | 0.150 | 1.20 | 20.70 | P68 | F17 | 0.750 | 0.60 | 27.00 |
| P07 | F02 | 0.600 | 5.25 | 82.80 | P38 | F09 | 0.150 | 1.20 | 20.70 | P69 | F17 | 0.750 | 0.90 | 27.00 |
| P08 | F02 | 0.600 | 5.25 | 82.80 | P39 | F10 | 0.125 | 1.80 | 11.25 | P70 | F17 | 0.750 | 0.60 | 27.00 |
| P09 | F03 | 0.600 | 6.00 | 82.80 | P40 | F10 | 0.125 | 1.80 | 11.25 | P71 | F17 | 0.750 | 0.60 | 27.00 |
| P10 | F03 | 0.600 | 5.25 | 82.80 | P41 | F10 | 0.125 | 1.80 | 11.25 | P72 | F18 | 1.000 | 0.75 | 33.00 |
| P11 | F03 | 0.600 | 5.25 | 82.80 | P42 | F11 | 1.000 | 3.30 | 33.00 | P73 | F18 | 1.000 | 0.75 | 33.00 |
| P12 | F04 | 0.600 | 8.25 | 82.80 | P43 | F11 | 1.000 | 4.20 | 33.00 | P74 | F19 | 1.000 | 0.45 | 33.00 |
| P13 | F04 | 0.600 | 6.75 | 82.80 | P44 | F11 | 1.000 | 4.20 | 33.00 | P75 | F19 | 1.000 | 0.75 | 33.00 |
| P14 | F04 | 0.600 | 6.00 | 82.80 | P45 | F12 | 1.000 | 4.20 | 33.00 | P76 | F19 | 1.000 | 0.75 | 33.00 |
| P15 | F04 | 0.600 | 5.25 | 82.80 | P46 | F12 | 1.000 | 3.00 | 33.00 | P77 | F20 | 5.000 | 3.30 | 80.00 |
| P16 | F04 | 0.600 | 5.25 | 82.80 | P47 | F12 | 1.000 | 3.00 | 33.00 | P78 | F20 | 5.000 | 3.60 | 80.00 |
| P17 | F05 | 0.600 | 9.00 | 82.80 | P48 | F13 | 1.000 | 2.55 | 33.00 | P79 | F20 | 5.000 | 3.60 | 80.00 |
| P18 | F05 | 0.600 | 6.75 | 82.80 | P49 | F13 | 1.000 | 2.10 | 33.00 | P80 | F21 | 30.000 | 0.45 | 90.00 |
| P19 | F05 | 0.600 | 7.50 | 82.80 | P50 | F13 | 1.000 | 2.10 | 33.00 | P81 | F21 | 30.000 | 0.75 | 90.00 |
| P20 | F05 | 0.600 | 6.00 | 82.80 | P51 | F14 | 1.000 | 3.60 | 33.00 | P82 | F21 | 30.000 | 0.75 | 90.00 |
| P21 | F05 | 0.200 | 5.70 | 27.60 | P52 | F14 | 1.000 | 3.60 | 33.00 | P83 | F22 | 10.000 | 0.75 | 100.00 |
| P22 | F05 | 0.200 | 5.70 | 27.60 | P53 | F15 | 0.500 | 3.60 | 28.00 | P84 | F22 | 10.000 | 0.45 | 100.00 |
| P23 | F06 | 0.400 | 0.75 | 55.20 | P54 | F15 | 0.500 | 3.60 | 28.00 | P85 | F22 | 5.000 | 0.30 | 80.00 |

(continued)

**Table B.1** (continued)

| Product | Family | Weight (kg) | Inv. cost (€) | Min. run (kg) | Product | Family | Weight (kg) | Inv. cost (€) | Min. run (kg) | Product | Family | Weight (kg) | Inv. cost (€) | Min. run (kg) |
|---|---|---|---|---|---|---|---|---|---|---|---|---|---|---|
| P24 | F06 | 0.200 | 0.75 | 27.60 | P55 | F16 | 0.500 | 1.80 | 16.50 | P86 | F22 | 5.000 | 0.45 | 80.00 |
| P25 | F06 | 0.200 | 0.75 | 27.60 | P56 | F16 | 0.500 | 1.80 | 16.50 | P87 | F22 | 5.000 | 0.75 | 80.00 |
| P26 | F07 | 0.380 | 0.75 | 19.76 | P57 | F16 | 0.500 | 1.80 | 16.50 | P88 | F22 | 5.000 | 0.45 | 80.00 |
| P27 | F07 | 0.380 | 0.75 | 19.76 | P58 | F16 | 0.500 | 1.80 | 16.50 | P89 | F22 | 5.000 | 0.45 | 80.00 |
| P28 | F08 | 0.400 | 1.35 | 55.20 | P59 | F16 | 0.500 | 1.80 | 16.50 | P90 | F23 | 0.150 | 1.80 | 32.30 |
| P29 | F08 | 0.400 | 1.35 | 55.20 | P60 | F16 | 0.500 | 1.80 | 16.50 | P91 | F23 | 0.150 | 1.80 | 32.30 |
| P30 | F08 | 0.600 | 1.65 | 82.80 | P61 | F17 | 0.750 | 0.90 | 27.00 | P92 | F23 | 0.150 | 1.80 | 32.30 |
| P31 | F08 | 0.600 | 1.65 | 82.80 | P62 | F17 | 0.750 | 0.90 | 27.00 | P93 | F23 | 0.150 | 1.80 | 32.30 |

**Table B.2** Products packing rates $p_j$ (kg/h)

| Product | J1 | J2 | J3 | J4 |
|---|---|---|---|---|
| P01 | | | 1710 | |
| P02 | | | 1710 | |
| P03 | | | 1710 | |
| P04 | | | 1710 | |
| P05 | | | 1710 | |
| P06 | | | 1710 | |
| P07 | | | 1710 | |
| P08 | | | 1710 | |
| P09 | | | 1710 | |
| P10 | | | 1710 | |
| P11 | | | 1710 | |
| P12 | | | 1710 | |
| P13 | | | 1710 | |
| P14 | | | 1710 | |
| P15 | | | 1710 | |
| P16 | | | 1710 | |
| P17 | | | 1710 | |
| P18 | | | 1710 | |
| F19 | | | 1710 | |
| P20 | | | 1710 | |
| P21 | | | 1820 | |
| P22 | | | 1820 | |
| P23 | | | 1320 | 1150 |
| P24 | | | 1390 | 1204 |

| Product | J1 | J2 | J3 | J4 |
|---|---|---|---|---|
| P32 | | | 2250 | 2520 |
| P33 | | | 2250 | 2520 |
| P34 | | | 1215 | 1215 |
| P35 | | | 1215 | 1215 |
| P36 | | | 1350 | 1350 |
| P37 | | | 1350 | 1350 |
| P38 | | | 1350 | 1350 |
| P39 | | | 850 | 850 |
| P40 | | | 850 | 850 |
| P41 | | | 850 | 850 |
| P42 | | 2200 | | |
| P43 | | 2200 | | |
| P44 | | 2200 | | |
| P45 | | 2200 | | |
| P46 | | 2200 | | |
| P47 | | 2200 | | |
| P48 | | 2200 | | |
| P49 | | 2200 | | |
| P50 | | 2200 | | |
| P51 | | 2200 | | |
| P52 | | 2200 | | |
| P53 | | 1150 | | |
| P54 | | 1150 | | |
| P55 | | 1150 | | |

| Product | J1 | J2 | J3 | J4 |
|---|---|---|---|---|
| P63 | | 1320 | | |
| P64 | | 1320 | | |
| P65 | | 1320 | | |
| P66 | | 1320 | | |
| P67 | | 1320 | | |
| P68 | | 1320 | | |
| P69 | | 1320 | | |
| P70 | | 1320 | | |
| P71 | | 1320 | | |
| P72 | | 2100 | | |
| P73 | | 2100 | | |
| P74 | | 2100 | | |
| P75 | | 2100 | | |
| P76 | | 2100 | | |
| P77 | 2250 | | | |
| P78 | 2100 | | | |
| P79 | 2100 | | | |
| P80 | 2790 | | | |
| P81 | 2790 | | | |
| P82 | 2790 | | | |
| P83 | 2350 | | | |
| P84 | 2350 | | | |
| P85 | 2250 | | | |
| P86 | 2250 | | | |

(continued)

**Table B.2** (continued)

| Product | J1 | J2 | J3 | J4 | Product | J1 | J2 | J3 | J4 | Product | J1 | J2 | J3 | J4 |
|---|---|---|---|---|---|---|---|---|---|---|---|---|---|---|
| P25 | | | 1390 | 1204 | P56 | | 1150 | | | P87 | 2150 | | | |
| P26 | | | 1140 | | P57 | | 1150 | | | P88 | 2150 | | | |
| P27 | | | 1140 | | P58 | | 1150 | | | P89 | 2150 | | | |
| P28 | | | 2400 | 2700 | P59 | | 1150 | | | P90 | | | | 690 |
| P29 | | | 2400 | 2700 | P60 | | 1150 | | | P91 | | | | 690 |
| P30 | | | 2250 | 2520 | P61 | | 1320 | | | P92 | | | | 690 |
| P31 | | | 2250 | 2520 | P62 | | 1320 | | | P93 | | | | 690 |

**Table B.3** Changeover times among families (h)

| Unit | Family | F01 | F02 | F03 | F04 | F05 | F06 | F07 | F08 | F09 | F10 | F11 | F12 | F13 | F14 | F15 | F16 | F17 | F18 | F20 | F19 | F21 | F22 | F23 |
|---|---|---|---|---|---|---|---|---|---|---|---|---|---|---|---|---|---|---|---|---|---|---|---|---|
| J3 | F01 | - | 0.00 | 0.00 | 0.50 | 0.50 | 0.50 | 0.50 | 1.00 | 1.50 | .50 | - | - | - | - | - | - | - | - | - | - | - | - | - |
| J3 | F02 | FS | - | 0.00 | 0.50 | 0.50 | 0.50 | 0.50 | 1.00 | 1.50 | 1.50 | - | - | - | - | - | - | - | - | - | - | - | - | - |
| J3 | F03 | FS | FS | - | 0.50 | 0.50 | 0.50 | 0.50 | 1.00 | 1.50 | 1.50 | - | - | - | - | - | - | - | - | - | - | - | - | - |
| J3 | F04 | FS | FS | FS | - | FS | 1.00 | 1.50 | 0.50 | 0.75 | 0.75 | - | - | - | - | - | - | - | - | - | - | - | - | - |
| J3 | F05 | FS | FS | FS | 0.25 | - | 0.50 | 0.75 | 1.00 | 1.50 | 1.50 | - | - | - | - | - | - | - | - | - | - | - | - | - |
| J3 | F06 | - | - | FS | FS | FS | - | 0.50 | FS | FS | FS | - | - | - | - | - | - | - | - | - | - | - | - | - |
| J4 | F06 | FS | - | - | - | - | - | - | FS | FS | FS | - | - | - | - | - | - | - | - | - | - | - | - | - |
| J3 | F07 | FS | FS | FS | FS | FS | FS | - | FS | FS | FS | - | - | - | - | - | - | - | - | - | - | - | - | - |
| J3 | F08 | FS | FS | FS | FS | FS | 2.00 | 2.00 | - | 0.50 | 0.50 | - | - | - | - | - | - | - | - | - | - | - | - | 2.00 |
| J4 | F08 | - | - | - | - | - | 2.00 | - | - | 0.50 | 0.50 | - | - | - | - | - | - | - | - | - | - | - | - | - |
| J3 | F09 | FS | FS | FS | FS | FS | 2.00 | 2.00 | FS | - | 0.50 | - | - | - | - | - | - | - | - | - | - | - | - | 2.00 |
| J4 | F09 | - | - | - | - | - | 2.00 | - | FS | - | 0.50 | - | - | - | - | - | - | - | - | - | - | - | - | - |
| J3 | F10 | FS | FS | FS | FS | FS | 2.00 | 2.00 | FS | FS | - | - | - | - | - | - | - | - | - | - | - | - | - | 2.00 |
| J4 | F10 | - | - | - | - | - | 2.00 | - | FS | FS | - | - | - | - | - | - | - | - | - | - | - | - | - | - |
| J2 | F11 | - | - | - | - | - | - | - | - | - | - | - | FS | 1.00 | 1.00 | 1.50 | 2.00 | 2.00 | 1.00 | 1.00 | - | - | - | - |
| J2 | F12 | - | - | - | - | - | - | - | - | - | - | 0.50 | - | 0.50 | 0.50 | 1.00 | 2.00 | 2.00 | 0.50 | 0.50 | - | - | - | - |
| J2 | F13 | - | - | - | - | - | - | - | - | - | - | FS | FS | - | 0.50 | 1.00 | 2.00 | 2.00 | FS | FS | - | - | - | - |
| J2 | F14 | - | - | - | - | - | - | - | - | - | - | FS | FS | FS | - | 0.50 | 2.00 | 2.00 | FS | FS | - | - | - | - |
| J2 | F15 | - | - | - | - | - | - | - | - | - | - | FS | FS | FS | FS | - | 1.50 | 2.00 | FS | FS | - | - | - | - |
| J2 | F16 | - | - | - | - | - | - | - | - | - | - | FS | FS | FS | FS | FS | - | 0.50 | FS | FS | - | - | - | - |
| J2 | F17 | - | - | - | - | - | - | - | - | - | - | FS | FS | 0.50 | 0.50 | 1.00 | 2.00 | - | FS | FS | - | - | - | - |
| J2 | F18 | - | - | - | - | - | - | - | - | - | - | FS | FS | 0.50 | 0.50 | 1.00 | 2.00 | 2.00 | - | FS | - | - | - | - |
| J3 | F19 | - | - | - | - | - | - | - | - | - | - | - | - | - | - | - | - | - | - | - | - | - | - | - |
| J1 | F20 | - | - | - | - | - | - | - | - | - | - | - | - | - | - | - | - | - | - | - | - | 0.50 | 1.00 | - |
| J1 | F21 | - | - | - | - | - | - | - | - | - | - | - | - | - | - | - | - | - | - | 0.50 | FS | - | 0.50 | - |
| J1 | F22 | - | - | - | - | - | - | - | - | - | - | - | - | - | - | - | - | - | - | - | FS | FS | - | - |
| J4 | F23 | - | - | - | - | - | FS | - | FS | FS | FS | - | - | - | - | - | - | - | - | - | FS | - | - | - |

Impossible processing sequence. FS: Forbidden processing sequence

**Table B.4** Correlation between changeover times and changeover costs

| Changeover time (h) | Changeover cost (€) |
|---|---|
| 0.00 | 750 |
| 0.25 | 1125 |
| 0.50 | 1800 |
| 0.75 | 2250 |
| 1.00 | 6000 |
| 1.50 | 15,000 |
| 2.00 | 22,000 |

**Table B.5** Production targets $\zeta_{pn}^{cup}$ (cups)

| Product | n1 | n2 | n3 | n4 | n5 | Product | n1 | n2 | n3 | n4 | n5 | Product | n1 | n2 | n3 | n4 | n5 |
|---|---|---|---|---|---|---|---|---|---|---|---|---|---|---|---|---|---|
| P01 | 3915 | | | | | P32 | | | | | 8249 | P63 | | 5753 | | | |
| P02 | | 2190 | | | | P33 | | | | | 4057 | P64 | | 1919 | | | |
| P03 | | | | 4416 | | P34 | | 2472 | | | | P65 | | 3648 | | | |
| P04 | | 6130 | | | 14,001 | P35 | | 2472 | | | | P66 | | | | | 1962 |
| P05 | | | | | 5480 | P36 | 26,496 | | | | | P67 | | 2648 | | | |
| P06 | | | | 5888 | | P37 | 9531 | | | | | P68 | | | | | 4683 |
| P07 | | | | | 11,241 | P38 | 4717 | | | | | P69 | | 2296 | | | |
| P08 | | | | | 4000 | P39 | | | | | 4093 | P70 | | | | | 544 |
| P09 | | | | 1888 | | P40 | | | | | 5743 | P71 | | | | | 219 |
| P10 | | | | | 1229 | P41 | | | | | 1172 | P72 | 2199 | | | 2040 | |
| P11 | | | | | 715 | P42 | | | | 3300 | | P73 | 1283 | | | | |
| P12 | 3560 | | | | | P43 | | | | | 1807 | P74 | | | | | 40 |
| P13 | | 4215 | | | | P44 | | | | | 2019 | P75 | | | | | 1071 |
| P14 | | | | 4416 | | P45 | | | | | 1578 | P76 | | | | | 117 |
| P15 | | | | | 6341 | P46 | | | | | 2518 | P77 | | | | | 57 |
| P16 | | | | | 3715 | P47 | | | | | 1690 | P78 | | | | | 1348 |
| P17 | 2319 | | | | | P48 | | | | | 2132 | P79 | | | | | 195 |
| P18 | | | | 2592 | | P49 | | | | | 5495 | P80 | | | 960 | | |
| P19 | | | 6138 | | | P50 | | | | | 1830 | P81 | | | 1160 | | |
| P20 | | | | | 6480 | P51 | | | | | 9380 | P82 | | | 900 | | |
| P21 | | | | | 1620 | P52 | | | | | 1272 | P83 | | | | | 710 |
| P22 | | | | | 1380 | P53 | | | | | 4386 | P84 | | | | | 290 |
| P23 | | | | 17,318 | | P54 | | | | | 1315 | P85 | | | | | 200 |

(continued)

**Table B.5** (continued)

| Product | n1 | n2 | n3 | n4 | n5 |
|---|---|---|---|---|---|
| P24 | | | | 4193 | |
| P25 | | | | 14,974 | |
| P26 | | 4671 | | | |
| P27 | | 1325 | | | |
| P28 | | 3312 | | | |
| P29 | | 3312 | | | |
| P30 | | | | | 3682 |
| P31 | | | | | 4801 |

| Product | n1 | n2 | n3 | n4 | n5 |
|---|---|---|---|---|---|
| P55 | | | | 4782 | |
| P56 | | | | 4316 | |
| P57 | | | | 3162 | |
| P58 | | | | 3188 | |
| P59 | | | | 3188 | |
| P60 | | | | 2316 | |
| P61 | | | | | 1.408 |
| P62 | | | | | 1262 |

| Product | n1 | n2 | n3 | n4 | n5 |
|---|---|---|---|---|---|
| P86 | | | | | 518 |
| P87 | | | | | 1130 |
| P88 | | | | | 1442 |
| P89 | | | | 3150 | |
| P90 | | 3140 | | | 10,140 |
| P91 | | 3120 | | | 5410 |
| P92 | | 9890 | | | 2900 |
| P93 | | 8220 | | | 1200 |

# Appendix C
# Detailed Production Plans for the Resource-Constrained Yogurt Production Process Case Studies

*Abstract* This appendix contains the detailed production plans for the yogurt production process case studies solved. More specifically, the detailed production plan for each product for Case Study I is reported in Table C.1. Table C.2 presents the detailed production plan for every product for Case Study II, and Table C.3 contains the detailed production plan for Case Study III.

**Table C.1** Case Study I. Detailed production plan (kg)

| Product | Unit | Monday | Tuesday | Wednesday | Thursday | Friday |
|---------|------|--------|---------|-----------|----------|--------|
| P44 | J2 | 0 | 0 | 0 | 1121 | 898 |
| P45 | J2 | 0 | 0 | 0 | 0 | 1578 |
| P46 | J2 | 0 | 0 | 0 | 0 | 2518 |
| P47 | J2 | 0 | 0 | 0 | 0 | 1690 |
| P48 | J2 | 0 | 0 | 0 | 0 | 2132 |
| P49 | J2 | 0 | 0 | 0 | 0 | 5495 |
| P50 | J2 | 0 | 0 | 0 | 0 | 1830 |
| P51 | J2 | 0 | 0 | 0 | 1100 | 8280 |
| P52 | J2 | 0 | 0 | 0 | 0 | 1272 |
| P53 | J2 | 0 | 0 | 0 | 2193 | 0 |
| P54 | J2 | 0 | 0 | 0 | 658 | 0 |
| P55 | J2 | 0 | 0 | 2351 | 40 | 0 |
| P56 | J2 | 0 | 0 | 0 | 2581 | 0 |
| P57 | J2 | 0 | 0 | 0 | 1581 | 0 |
| P58 | J2 | 0 | 0 | 0 | 1594 | 0 |
| P59 | J2 | 0 | 0 | 1578 | 17 | 0 |
| P60 | J2 | 0 | 0 | 1158 | 0 | 0 |

(continued)

G. M. Kopanos and L. Puigjaner, *Solving Large-Scale Production Scheduling and Planning in the Process Industries*,
https://doi.org/10.1007/978-3-030-01183-3

**Table C.1** (continued)

| Product | Unit | Monday | Tuesday | Wednesday | Thursday | Friday |
|---|---|---|---|---|---|---|
| P61 | J2 | 0 | 0 | 1056 | 0 | 0 |
| P62 | J2 | 0 | 0 | 947 | 0 | 0 |
| P63 | J2 | 0 | 4315 | 0 | 0 | 0 |
| P64 | J2 | 0 | 1439 | 0 | 0 | 0 |
| P65 | J2 | 0 | 2736 | 0 | 0 | 0 |
| P66 | J2 | 0 | 27 | 1445 | 0 | 0 |
| P67 | J2 | 0 | 1986 | 0 | 0 | 0 |
| P68 | J2 | 0 | 27 | 3485 | 0 | 0 |
| P69 | J2 | 0 | 1722 | 0 | 0 | 0 |
| P70 | J2 | 0 | 336 | 72 | 0 | 0 |
| P71 | J2 | 0 | 137 | 27 | 0 | 0 |
| P72 | J2 | 4239 | 0 | 0 | 0 | 0 |
| P73 | J2 | 1283 | 0 | 0 | 0 | 0 |
| P74 | J2 | 0 | 0 | 0 | 40 | 0 |
| P75 | J2 | 0 | 0 | 0 | 1071 | 0 |
| P76 | J2 | 0 | 0 | 0 | 117 | 0 |
| P77 | J1 | 0 | 0 | 0 | 0 | 285 |
| P78 | J1 | 0 | 0 | 0 | 0 | 6740 |
| P79 | J1 | 0 | 0 | 0 | 0 | 975 |
| P80 | J1 | 25,035 | 3765 | 0 | 0 | 0 |
| P81 | J1 | 0 | 29,018 | 5783 | 0 | 0 |
| P82 | J1 | 0 | 0 | 27,000 | 0 | 0 |
| P83 | J1 | 0 | 0 | 0 | 0 | 7100 |
| P84 | J1 | 0 | 0 | 0 | 0 | 2900 |
| P85 | J1 | 0 | 0 | 0 | 1000 | 0 |
| P86 | J1 | 0 | 0 | 0 | 0 | 2590 |
| P87 | J1 | 0 | 0 | 0 | 0 | 5650 |
| P88 | J1 | 0 | 0 | 0 | 5242 | 1968 |
| P89 | J1 | 0 | 0 | 0 | 15,750 | 0 |
| P90 | J4 | 0 | 471 | 1521 | 0 | 0 |
| P91 | J4 | 0 | 468 | 812 | 0 | 0 |
| P92 | J4 | 0 | 1484 | 435 | 0 | 0 |
| P93 | J4 | 0 | 1233 | 180 | 0 | 0 |
| Total daily production | | 47,708 | 59,333 | 51,531 | 62,764 | 97,916 |

**Table C.2** Case Study II. Detailed production plan (kg)

| Product | Unit | Wednesday | Thursday | Friday | Product | Unit | Wednesday | Thursday | Friday |
|---|---|---|---|---|---|---|---|---|---|
| P01 | J3 | 0 | 0 | 0 | P26 | J3 | 0 | 0 | 0 |
| P02 | J3 | 0 | 0 | 0 | P27 | J3 | 0 | 0 | 0 |
| P03 | J3 | 0 | 2650 | 0 | P28 | J3 | 0 | 0 | 0 |
| P04 | J3 | 0 | 0 | 8401 | P28 | J4 | 0 | 0 | 0 |
| P05 | J3 | 0 | 0 | 3288 | P29 | J3 | 0 | 0 | 0 |
| P06 | J3 | 0 | 3533 | 0 | P29 | J4 | 0 | 0 | 0 |
| P07 | J3 | 0 | 0 | 6745 | P30 | J3 | 0 | 83 | 0 |
| P08 | J3 | 0 | 0 | 2400 | P30 | J4 | 0 | 0 | 2126 |
| P09 | J3 | 0 | 1133 | 0 | P31 | J3 | 0 | 0 | 0 |
| P10 | J3 | 0 | 737 | 0 | P31 | J4 | 0 | 0 | 2881 |
| P11 | J3 | 0 | 429 | 0 | P32 | J3 | 0 | 4949 | 0 |
| P12 | J3 | 0 | 0 | 0 | P32 | J4 | 0 | 0 | 0 |
| P13 | J3 | 0 | 0 | 0 | P33 | J3 | 0 | 555 | 0 |
| P14 | J3 | 0 | 2650 | 0 | P33 | J4 | 0 | 0 | 1879 |
| P15 | J3 | 0 | 2676 | 1129 | P34 | J3 | 0 | 0 | 0 |
| P16 | J3 | 0 | 0 | 2229 | P34 | J4 | 0 | 0 | 0 |
| P17 | J3 | 0 | 0 | 0 | P35 | J3 | 0 | 0 | 0 |
| P18 | J3 | 0 | 1555 | 0 | P35 | J4 | 0 | 0 | 0 |
| P19 | J3 | 3683 | 0 | 0 | P36 | J3 | 0 | 0 | 0 |
| P20 | J3 | 0 | 0 | 3888 | P36 | J4 | 0 | 0 | 0 |
| P21 | J3 | 0 | 0 | 324 | P37 | J3 | 0 | 0 | 0 |
| P22 | J3 | 0 | 0 | 276 | P37 | J4 | 0 | 0 | 0 |
| P23 | J3 | 0 | 0 | 0 | P38 | J3 | 0 | 0 | 0 |
| P23 | J4 | 6267 | 0 | 0 | P38 | J4 | 0 | 0 | 0 |

(continued)

**Table C.2** (continued)

| Product | Unit | Wednesday | Thursday | Friday |
|---|---|---|---|---|
| P24 | J3 | 0 | 0 | 0 |
| P24 | J4 | 839 | 0 | 0 |
| P25 | J3 | 0 | 0 | 0 |
| P25 | J4 | 2995 | 0 | 0 |

| Product | Unit | Wednesday | Thursday | Friday |
|---|---|---|---|---|
| P41 | J3 | 0 | 147 | 0 |
| P41 | J4 | 0 | 0 | 0 |
| P42 | J2 | 0 | 3300 | 0 |
| P43 | J2 | 0 | 33 | 1774 |
| P44 | J2 | 0 | 1088 | 931 |
| P45 | J2 | 0 | 0 | 1578 |
| P46 | J2 | 0 | 0 | 2518 |
| P47 | J2 | 0 | 0 | 1690 |
| P48 | J2 | 0 | 0 | 2132 |
| P49 | J2 | 0 | 0 | 5495 |
| P50 | J2 | 0 | 0 | 1830 |
| P51 | J2 | 0 | 1100 | 8280 |
| P52 | J2 | 0 | 0 | 1272 |
| P53 | J2 | 0 | 2193 | 0 |
| P54 | J2 | 0 | 658 | 0 |
| P55 | J2 | 2391 | 0 | 0 |
| P56 | J2 | 17 | 2142 | 0 |
| P57 | J2 | 17 | 1565 | 0 |
| P58 | J2 | 17 | 1578 | 0 |

| Product | Unit | Wednesday | Thursday | Friday |
|---|---|---|---|---|
| P39 | J3 | 0 | 512 | 0 |
| P39 | J4 | 0 | 0 | 0 |
| P40 | J3 | 0 | 718 | 0 |
| P40 | J4 | 0 | 0 | 0 |

| Product | Unit | Wednesday | Thursday | Friday |
|---|---|---|---|---|
| P68 | J2 | 3485 | 0 | 0 |
| P69 | J2 | 0 | 0 | 0 |
| P70 | J2 | 72 | 0 | 0 |
| P71 | J2 | 27 | 0 | 0 |
| P72 | J2 | 0 | 0 | 0 |
| P73 | J2 | 0 | 0 | 0 |
| P74 | J2 | 0 | 40 | 0 |
| P75 | J2 | 0 | 1071 | 0 |
| P76 | J2 | 0 | 117 | 0 |
| P77 | J1 | 0 | 0 | 285 |
| P78 | J1 | 0 | 0 | 6740 |
| P79 | J1 | 0 | 0 | 975 |
| P80 | J1 | 0 | 0 | 0 |
| P81 | J1 | 5783 | 0 | 0 |
| P82 | J1 | 27,000 | 0 | 0 |
| P83 | J1 | 0 | 0 | 7100 |
| P84 | J1 | 0 | 0 | 2900 |
| P85 | J1 | 0 | 1000 | 0 |
| P86 | J1 | 0 | 1410 | 1180 |

(continued)

**Table C.2** (continued)

| Product | Unit | Wednesday | Thursday | Friday | Product | Unit | Wednesday | Thursday | Friday |
|---------|------|-----------|----------|--------|---------|------|-----------|----------|--------|
| P59 | J2 | 1504 | 90 | 0 | P87 | J1 | 0 | 0 | 5650 |
| P60 | J2 | 1142 | 17 | 0 | P88 | J1 | 0 | 7210 | 0 |
| P61 | J2 | 1056 | 0 | 0 | P89 | J1 | 0 | 15,750 | 0 |
| P62 | J2 | 947 | 0 | 0 | P90 | J4 | 0 | 1521 | 0 |
| P63 | J2 | 0 | 0 | 0 | P91 | J4 | 0 | 812 | 0 |
| P64 | J2 | 0 | 0 | 0 | P92 | J4 | 0 | 435 | 0 |
| P65 | J2 | 0 | 0 | 0 | P93 | J4 | 0 | 180 | 0 |
| P66 | J2 | 1445 | 0 | 0 | | | | | |
| P67 | J2 | 0 | 0 | 0 | Total daily production | | 58,684 | 65,632 | 87,895 |

**Table C.3** Case Study III. Detailed production plan (kg)

| Product | Unit | Thursday | Friday | Product | Unit | Thursday | Friday | Product | Unit | Thursday | Friday | Product | Unit | Thursday | Friday |
|---|---|---|---|---|---|---|---|---|---|---|---|---|---|---|---|
| P01 | J3 | 841 | 0 | P26 | J3 | 0 | 0 | P41 | J3 | 0 | 0 | P68 | J2 | 0 | 0 |
| P02 | J3 | 0 | 0 | P27 | J3 | 0 | 0 | P41 | J4 | 0 | 285 | P69 | J2 | 0 | 0 |
| P03 | J3 | 2650 | 0 | P28 | J3 | 0 | 0 | P42 | J2 | 3,300 | 0 | P70 | J2 | 0 | 0 |
| P04 | J3 | 0 | 6884 | P28 | J4 | 0 | 858 | P43 | J2 | 1807 | 0 | P71 | J2 | 0 | 0 |
| P05 | J3 | 0 | 4791 | P29 | J3 | 0 | 0 | P44 | J2 | 1626 | 0 | P72 | J2 | 0 | 0 |
| P06 | J3 | 3533 | 0 | P29 | J4 | 0 | 0 | P45 | J2 | 0 | 1578 | P73 | J2 | 0 | 0 |
| P07 | J3 | 974 | 5771 | P30 | J3 | 0 | 0 | P46 | J2 | 0 | 2518 | P74 | J2 | 0 | 40 |
| P08 | J3 | 0 | 0 | P30 | J4 | 0 | 2209 | P47 | J2 | 0 | 1690 | P75 | J2 | 0 | 1071 |
| P09 | J3 | 1133 | 0 | P31 | J3 | 0 | 0 | P48 | J2 | 0 | 2132 | P76 | J2 | 0 | 117 |
| P10 | J3 | 737 | 0 | P31 | J4 | 0 | 2881 | P49 | J2 | 0 | 5495 | P77 | J1 | 285 | 0 |
| P11 | J3 | 568 | 0 | P32 | J3 | 0 | 0 | P50 | J2 | 0 | 1830 | P78 | J1 | 3515 | 3225 |
| P12 | J3 | 0 | 0 | P32 | J4 | 0 | 4949 | P51 | J2 | 33 | 9347 | P79 | J1 | 0 | 975 |
| P13 | J3 | 977 | 0 | P33 | J3 | 0 | 0 | P52 | J2 | 1067 | 205 | P80 | J1 | 1395 | 2205 |
| P14 | J3 | 2650 | 0 | P33 | J4 | 0 | 1801 | P53 | J2 | 2193 | 0 | P81 | J1 | 0 | 7200 |
| P15 | J3 | 0 | 3805 | P34 | J3 | 0 | 0 | P54 | J2 | 54 | 604 | P82 | J1 | 0 | 0 |
| P16 | J3 | 0 | 2469 | P34 | J4 | 0 | 0 | P55 | J2 | 40 | 0 | P83 | J1 | 0 | 7100 |
| P17 | J3 | 1157 | 0 | P35 | J3 | 0 | 0 | P56 | J2 | 2158 | 0 | P84 | J1 | 0 | 2900 |
| P18 | J3 | 1555 | 0 | P35 | J4 | 0 | 0 | P57 | J2 | 1581 | 0 | P85 | J1 | 1000 | 0 |
| P19 | J3 | 0 | 0 | P36 | J3 | 0 | 0 | P58 | J2 | 993 | 0 | P86 | J1 | 0 | 2590 |
| P20 | J3 | 0 | 3888 | P36 | J4 | 0 | 0 | P59 | J2 | 17 | 0 | P87 | J1 | 0 | 2363 |
| P21 | J3 | 0 | 444 | P37 | J3 | 0 | 0 | P60 | J2 | 0 | 0 | P88 | J1 | 5332 | 1878 |
| P22 | J3 | 0 | 358 | P37 | J4 | 0 | 0 | P61 | J2 | 0 | 0 | P89 | J1 | 15,750 | 0 |
| P23 | J4 | 6267 | 0 | P38 | J3 | 0 | 0 | P62 | J2 | 0 | 0 | P90 | J4 | 0 | 0 |
| P24 | J3 | 0 | 0 | P38 | J4 | 0 | 0 | P63 | J2 | 0 | 0 | P91 | J4 | 0 | 0 |
| P24 | J4 | 0 | 0 | P39 | J3 | 0 | 0 | P64 | J2 | 1392 | 0 | P92 | J4 | 0 | 0 |
| P25 | J3 | 0 | 0 | P39 | J4 | 0 | 512 | P65 | J2 | 0 | 0 | P93 | J4 | 0 | 0 |
| P25 | J4 | 2758 | 0 | P40 | J3 | 0 | 0 | P66 | J2 | 0 | 0 | | | | |
| | | | | P40 | J4 | 0 | 718 | P67 | J2 | 0 | 0 | Total daily production | | 69,336 | 99,685 |

# Appendix D
# MIP-Based Solution Strategy Pseudocodes

***Abstract*** This appendix presents some representative pseudocodes for the two-stage MIP-based solution strategy described in Chap. 8.

---

**Algorithm D.1**: Pseudocode for iterative procedure in Constructive Step

---

Set step = 1, initial = 1 & pos(i) parameter. Also, set $I^{in} = \emptyset$;

FOR $z$ = initial to card(i) by step
      LOOP i
            IF pos(i) ≤ z
                $I^{in}$=yes
            END IF
      END LOOP
      SOLVE MIP model
      fix $Y_{isj}$ & $X_{ii'j}$ binary variables $\forall i \in I^{in}$
END FOR

---

G. M. Kopanos and L. Puigjaner, *Solving Large-Scale Production Scheduling and Planning in the Process Industries*,
https://doi.org/10.1007/978-3-030-01183-3

279

**Algorithm D.2**: : Pseudocode for iterative procedure in Improvement Step

Refer to Méndez and Cerdá (2003a) for an explanation of parameter $n$, subsets $ISi$, $ISSii'$, and Reordering-MIP model.

Set $iter^{max}$ $n = 1$, order(i), reins=1 parameters & $IS_i$ subset
$fixY_{isj} = Y_{isj}$, $fixX_{ii'j} = X_{ii'j}$ (solucion of constructiove step)
iteration = 1

WHILE ($OF^{reins}$ is better than $OF^{reord}$ or iteration = 1)

   → *Reordering Stage*
   $Y_{isj} = fixY_{isj}$
   $iter = 1$

   WHILE ($OF_{iter}$ is better than $OF_{iter-1}$ and $iter \leq iter^{max}$)
      CLEAR $ISS_{ii'}$ subset
      asses $ISS_{ii'}$ subset
      CLEAR all variables apart from $Y_{isj}$
      SOLVE Reordering-MIP mode
      $iter = iter+1$
   END WHILE

   save save best solution of reordering stage:
   CLEAR $fixX_{ii'j}$, and set $fixX_{ii'j} = X_{ii'j}$ & $OF^{reord} = OF_{iter-1}$

   → *Reinsertion Stage*
   $iter = 1$
   FOR z = reins to card(i) by reis
      LOOP i
        IF order(i) $\leq z$ - reins + 1
           CLEAR all variables related to i (e.g., $Y_{isj}$, $X_{ii'j}$, $C_{is}$, etc.)
        ELSE
           $Y_{isj} = fixY_{isj}$ & $X_{ii'j} = fixX_{ii'j}$
        END IF
      END LOOP
      SOLVE MIP model
      IF $OF_{iter}$ is better than $OF^{reord}$
        Save Solution(iter) (e.g., save $OF$, $Y_{isj}$, $X_{ii'j}$, $C_{is}$, $C_{is}$, , etc.)
      END IF
      $iter = iter +1$
   END FOR

   save best solution of reinsertion stage:
   CLEAR $fixY_{isj}$, $fixX_{ii'j}$ & $OF^{reins}$
   $OF^{reins}$ is equal to the best $OF_{iter}$
   set $fixY_{isj} = Y_{isj}$, $fixX_{ii'j} = X_{ii'j}$ only for $OF_{iter} = OF^{reins}$

   iteration = iteration + 1

END WHILE

# Appendix E
# Data for the Pharmaceutical Multistage Batch Process

*Abstract* This appendix contains the data for the changeover times for stages S3–S6 and the processing times for the pharmaceutical process described and studied in Chap. 8 (Tables E.1, E.2, E.3 and E.4).

© Springer Nature Switzerland AG 2019         281
G. M. Kopanos and L. Puigjaner, *Solving Large-Scale Production Scheduling and Planning in the Process Industries*,
https://doi.org/10.1007/978-3-030-01183-3

**Table E.1** Changeover times in stage S3 (h)

|  | P01 | P08 | P12 | P13 | P16 | P17 | P20 | P21 | P23 | P24 | P26 | P27 |
|---|---|---|---|---|---|---|---|---|---|---|---|---|
| P01 | 0.00 | 1.80 | 1.80 | 1.35 | 1.80 | 1.35 | 1.35 | 1.80 | 1.80 | 1.35 | 1.80 | 1.80 |
| P08 | 1.80 | 0.00 | 1.35 | 1.80 | 1.35 | 1.80 | 1.80 | 1.80 | 1.35 | 1.80 | 1.35 | 1.80 |
| P12 | 1.80 | 1.35 | 0.00 | 1.80 | 1.35 | 1.80 | 1.80 | 1.80 | 1.35 | 1.80 | 1.35 | 1.80 |
| P13 | 1.35 | 1.80 | 1.80 | 0.00 | 1.80 | 1.35 | 1.35 | 1.80 | 1.80 | 1.35 | 1.80 | 1.80 |
| P16 | 1.80 | 1.35 | 1.35 | 1.80 | 0.00 | 1.80 | 1.80 | 1.80 | 1.35 | 1.80 | 1.35 | 1.80 |
| P17 | 1.35 | 1.80 | 1.80 | 1.35 | 1.80 | 0.00 | 1.35 | 1.80 | 1.80 | 1.35 | 1.80 | 1.80 |
| P20 | 1.35 | 1.80 | 1.80 | 1.35 | 1.80 | 1.35 | 0.00 | 1.80 | 1.80 | 1.35 | 1.80 | 1.80 |
| P21 | 1.80 | 1.80 | 1.80 | 1.80 | 1.80 | 1.80 | 1.80 | 0.00 | 1.80 | 1.80 | 1.80 | 1.35 |
| P23 | 1.80 | 1.35 | 1.35 | 1.80 | 1.35 | 1.80 | 1.80 | 1.80 | 0.00 | 1.80 | 1.35 | 1.80 |
| P24 | 1.35 | 1.80 | 1.80 | 1.35 | 1.80 | 1.35 | 1.35 | 1.80 | 1.80 | 0.00 | 1.80 | 1.80 |
| P26 | 1.80 | 1.35 | 1.35 | 1.80 | 1.35 | 1.80 | 1.80 | 1.80 | 1.35 | 1.80 | 0.00 | 1.80 |
| P27 | 1.80 | 1.80 | 1.80 | 1.80 | 1.80 | 1.80 | 1.80 | 1.35 | 1.80 | 1.80 | 1.80 | 0.00 |

**Table E.2** Changeover times in stages S4 and S5 (h)

|     | P01 | P02 | P03 | P04 | P05 | P06 | P07 | P08 | P09 | P10 | P11 | P12 | P13 | P14 | P15 |
|-----|-----|-----|-----|-----|-----|-----|-----|-----|-----|-----|-----|-----|-----|-----|-----|
| P01 | 0   | 0.9 | 1.8 | 1.8 | 1.8 | 1.8 | 0.9 | 1.8 | 1.8 | 1.8 | 1.8 | 1.8 | 0.9 | 0.9 | 1.8 |
| P02 | 0.9 | 0   | 1.8 | 1.8 | 1.8 | 1.8 | 0.9 | 1.8 | 1.8 | 1.8 | 1.8 | 1.8 | 0.9 | 0.9 | 1.8 |
| P03 | 1.8 | 1.8 | 0   | 1.8 | 0.9 | 1.8 | 1.8 | 1.8 | 1.8 | 0.9 | 0.9 | 1.8 | 1.8 | 1.8 | 0.9 |
| P04 | 1.8 | 1.8 | 1.8 | 0   | 1.8 | 0.9 | 1.8 | 0.9 | 0.9 | 1.8 | 1.8 | 0.9 | 1.8 | 1.8 | 1.8 |
| P05 | 1.8 | 1.8 | 0.9 | 1.8 | 0   | 1.8 | 1.8 | 1.8 | 1.8 | 0.9 | 0.9 | 0.9 | 1.8 | 1.8 | 0.9 |
| P06 | 1.8 | 1.8 | 1.8 | 1.8 | 1.8 | 0   | 1.8 | 0.9 | 0.9 | 1.8 | 1.8 | 1.8 | 1.8 | 1.8 | 1.8 |
| P07 | 0.9 | 0.9 | 1.8 | 1.8 | 1.8 | 1.8 | 0   | 1.8 | 1.8 | 1.8 | 1.8 | 0.9 | 0.9 | 0.9 | 1.8 |
| P08 | 1.8 | 0.9 | 1.8 | 0.9 | 1.8 | 0.9 | 1.8 | 0   | 0.9 | 1.8 | 1.8 | 0.9 | 1.8 | 1.8 | 1.8 |
| P09 | 1.8 | 0.9 | 1.8 | 0.9 | 1.8 | 0.9 | 1.8 | 1.8 | 0   | 1.8 | 1.8 | 0.9 | 1.8 | 1.8 | 1.8 |
| P10 | 1.8 | 0.9 | 0.9 | 1.8 | 0.9 | 0.9 | 1.8 | 1.8 | 1.8 | 0   | 0.9 | 1.8 | 1.8 | 1.8 | 0.9 |
| P11 | 1.8 | 0.9 | 0.9 | 1.8 | 0.9 | 1.8 | 1.8 | 1.8 | 1.8 | 0.9 | 0   | 1.8 | 1.8 | 1.8 | 0.9 |
| P12 | 1.8 | 0.9 | 1.8 | 0.9 | 1.8 | 0.9 | 1.8 | 0.9 | 0.9 | 1.8 | 1.8 | 0   | 1.8 | 1.8 | 1.8 |
| P13 | 0.9 | 0.9 | 1.8 | 1.8 | 1.8 | 1.8 | 0.9 | 1.8 | 1.8 | 1.8 | 1.8 | 1.8 | 0   | 0.9 | 1.8 |
| P14 | 0.9 | 0.9 | 1.8 | 1.8 | 1.8 | 1.8 | 0.9 | 1.8 | 1.8 | 1.8 | 1.8 | 1.8 | 0.9 | 0   | 1.8 |
| P15 | 1.8 | 1.8 | 0.9 | 1.8 | 0.9 | 1.8 | 1.8 | 1.8 | 1.8 | 0.9 | 0.9 | 1.8 | 1.8 | 1.8 | 0   |
| P16 | 1.8 | 1.8 | 1.8 | 0.9 | 1.8 | 0.9 | 1.8 | 0.9 | 0.9 | 1.8 | 1.8 | 0.9 | 1.8 | 1.8 | 1.8 |
| P17 | 0.9 | 0.9 | 1.8 | 1.8 | 1.8 | 1.8 | 0.9 | 1.8 | 1.8 | 1.8 | 1.8 | 1.8 | 0.9 | 0.9 | 1.8 |
| P18 | 0.9 | 0.9 | 1.8 | 1.8 | 1.8 | 1.8 | 0.9 | 1.8 | 1.8 | 1.8 | 1.8 | 1.8 | 0.9 | 0.9 | 1.8 |
| P19 | 1.8 | 1.8 | 1.8 | 0.9 | 1.8 | 0.9 | 1.8 | 0.9 | 0.9 | 1.8 | 1.8 | 0.9 | 1.8 | 1.8 | 1.8 |
| P20 | 0.9 | 1.8 | 1.8 | 1.8 | 1.8 | 1.8 | 0.9 | 1.8 | 1.8 | 1.8 | 1.8 | 1.8 | 0.9 | 0.9 | 1.8 |
| P21 | 1.8 | 1.8 | 0.9 | 1.8 | 0.9 | 1.8 | 1.8 | 1.8 | 1.8 | 0.9 | 0.9 | 1.8 | 1.8 | 1.8 | 0.9 |
| P22 | 1.8 | 1.8 | 0.9 | 1.8 | 0.9 | 1.8 | 1.8 | 1.8 | 1.8 | 0.9 | 0.9 | 1.8 | 1.8 | 1.8 | 0.9 |
| P23 | 1.8 | 1.8 | 1.8 | 0.9 | 1.8 | 0.9 | 1.8 | 0.9 | 0.9 | 1.8 | 1.8 | 0.9 | 1.8 | 1.8 | 1.8 |
| P24 | 0.9 | 0.9 | 1.8 | 1.8 | 1.8 | 1.8 | 0.9 | 1.8 | 1.8 | 1.8 | 1.8 | 1.8 | 0.9 | 0.9 | 1.8 |

(continued)

**Table E.2** (continued)

| | P01 | P02 | P03 | P04 | P05 | P06 | P07 | P08 | P09 | P10 | P11 | P12 | P13 | P14 | P15 |
|---|---|---|---|---|---|---|---|---|---|---|---|---|---|---|---|
| P25 | 0.9 | 0.9 | 1.8 | 1.8 | 1.8 | 1.8 | 0.9 | 1.8 | 1.8 | 1.8 | 1.8 | 1.8 | 0.9 | 0.9 | 1.8 |
| P26 | 1.8 | 1.8 | 1.8 | 0.9 | 1.8 | 0.9 | 1.8 | 0.9 | 0.9 | 1.8 | 1.8 | 0.9 | 1.8 | 1.8 | 1.8 |
| P27 | 1.8 | 1.8 | 0.9 | 1.8 | 0.9 | 1.8 | 1.8 | 1.8 | 1.8 | 0.9 | 0.9 | 1.8 | 1.8 | 1.8 | 0.9 |
| P28 | 1.8 | 1.8 | 0.9 | 1.8 | 0.9 | 1.8 | 1.8 | 1.8 | 1.8 | 0.9 | 0.9 | 1.8 | 1.8 | 1.8 | 0.9 |
| P29 | 0.9 | 0.9 | 1.8 | 1.8 | 1.8 | 1.8 | 0.9 | 1.8 | 1.8 | 1.8 | 1.8 | 1.8 | 0.9 | 0.9 | 1.8 |
| P30 | 1.8 | 1.8 | 0.9 | 1.8 | 0.9 | 1.8 | 1.8 | 1.8 | 1.8 | 0.9 | 0.9 | 1.8 | 1.8 | 1.8 | 0.9 |

| | P16 | P17 | P18 | P19 | P20 | P21 | P22 | P23 | P24 | P25 | P26 | P27 | P28 | P29 | P30 |
|---|---|---|---|---|---|---|---|---|---|---|---|---|---|---|---|
| P01 | 1.8 | 0.9 | 0.9 | 1.8 | 0.9 | 1.8 | 1.8 | 1.8 | 0.9 | 0.9 | 1.8 | 1.8 | 1.8 | 0.9 | 1.8 |
| P02 | 1.8 | 0.9 | 0.9 | 1.8 | 0.9 | 1.8 | 1.8 | 1.8 | 0.9 | 0.9 | 1.8 | 1.8 | 1.8 | 0.9 | 1.8 |
| P03 | 1.8 | 1.8 | 1.8 | 1.8 | 1.8 | 0.9 | 0.9 | 1.8 | 1.8 | 1.8 | 0.9 | 0.9 | 0.9 | 1.8 | 0.9 |
| P04 | 0.9 | 1.8 | 1.8 | 0.9 | 1.8 | 1.8 | 1.8 | 0.9 | 1.8 | 1.8 | 1.8 | 1.8 | 1.8 | 1.8 | 1.8 |
| P05 | 1.8 | 0.9 | 1.8 | 1.8 | 1.8 | 0.9 | 0.9 | 1.8 | 1.8 | 1.8 | 0.9 | 1.8 | 0.9 | 1.8 | 0.9 |
| P06 | 0.9 | 1.8 | 1.8 | 1.8 | 1.8 | 1.8 | 1.8 | 1.8 | 1.8 | 1.8 | 1.8 | 1.8 | 1.8 | 1.8 | 1.8 |
| P07 | 1.8 | 0.9 | 0.9 | 0.9 | 0.9 | 1.8 | 1.8 | 0.9 | 0.9 | 0.9 | 1.8 | 1.8 | 1.8 | 0.9 | 1.8 |
| P08 | 0.9 | 1.8 | 1.8 | 0.9 | 1.8 | 1.8 | 1.8 | 0.9 | 1.8 | 1.8 | 1.8 | 1.8 | 1.8 | 1.8 | 1.8 |
| P09 | 0.9 | 1.8 | 1.8 | 0.9 | 1.8 | 1.8 | 1.8 | 0.9 | 1.8 | 1.8 | 1.8 | 1.8 | 1.8 | 1.8 | 1.8 |
| P10 | 1.8 | 1.8 | 1.8 | 1.8 | 1.8 | 0.9 | 0.9 | 1.8 | 1.8 | 1.8 | 0.9 | 1.8 | 0.9 | 1.8 | 0.9 |
| P11 | 1.8 | 1.8 | 1.8 | 1.8 | 1.8 | 0.9 | 0.9 | 1.8 | 1.8 | 1.8 | 0.9 | 1.8 | 0.9 | 1.8 | 0.9 |
| P12 | 0.9 | 1.8 | 1.8 | 0.9 | 1.8 | 1.8 | 1.8 | 0.9 | 1.8 | 1.8 | 1.8 | 1.8 | 1.8 | 1.8 | 1.8 |
| P13 | 1.8 | 0.9 | 0.9 | 1.8 | 0.9 | 1.8 | 1.8 | 1.8 | 0.9 | 0.9 | 1.8 | 1.8 | 1.8 | 1.8 | 1.8 |
| P14 | 1.8 | 0.9 | 0.9 | 1.8 | 0.9 | 1.8 | 1.8 | 1.8 | 0.9 | 0.9 | 1.8 | 1.8 | 1.8 | 1.8 | 1.8 |
| P15 | 1.8 | 1.8 | 1.8 | 1.8 | 1.8 | 0.9 | 0.9 | 1.8 | 1.8 | 1.8 | 0.9 | 0.9 | 0.9 | 1.8 | 0.9 |
| P16 | 0 | 1.8 | 1.8 | 0.9 | 1.8 | 1.8 | 1.8 | 0.9 | 1.8 | 1.8 | 1.8 | 1.8 | 1.8 | 0.9 | 1.8 |
| P17 | 1.8 | 0 | 0.9 | 1.8 | 0.9 | 1.8 | 1.8 | 1.8 | 0.9 | 0.9 | 1.8 | 1.8 | 1.8 | 0.9 | 1.8 |

(continued)

**Table E.2** (continued)

| | P16 | P17 | P18 | P19 | P20 | P21 | P22 | P23 | P24 | P25 | P26 | P27 | P28 | P29 | P30 |
|-----|-----|-----|-----|-----|-----|-----|-----|-----|-----|-----|-----|-----|-----|-----|-----|
| P18 | 1.8 | 0.9 | 0 | 1.8 | 0.9 | 1.8 | 1.8 | 1.8 | 0.9 | 0.9 | 1.8 | 1.8 | 1.8 | 0.9 | 1.8 |
| P19 | 0.9 | 1.8 | 1.8 | 0 | 1.8 | 1.9 | 1.8 | 0.9 | 1.8 | 1.8 | 0.9 | 1.8 | 1.8 | 1.8 | 1.8 |
| P20 | 1.8 | 0.9 | 0.9 | 1.8 | 0 | 1.8 | 1.8 | 1.8 | 0.9 | 0.9 | 1.8 | 1.8 | 1.8 | 0.9 | 1.8 |
| P21 | 1.8 | 1.8 | 1.8 | 1.8 | 1.8 | 0 | 0.9 | 1.8 | 1.8 | 1.8 | 1.8 | 0.9 | 0.9 | 1.8 | 0.9 |
| P22 | 1.8 | 1.8 | 1.8 | 1.8 | 1.8 | 0.9 | 0 | 1.8 | 1.8 | 1.8 | 1.8 | 0.9 | 0.9 | 1.8 | 0.9 |
| P23 | 0.9 | 1.8 | 1.8 | 0.9 | 1.8 | 1.8 | 1.8 | 0 | 1.8 | 1.8 | 0.9 | 1.8 | 1.8 | 1.8 | 1.8 |
| P24 | 1.8 | 0.9 | 0.9 | 1.8 | 0.9 | 1.8 | 1.8 | 1.8 | 0 | 0.9 | 1.8 | 1.8 | 1.8 | 0.9 | 1.8 |
| P25 | 1.8 | 0.9 | 0.9 | 1.8 | 0.9 | 1.8 | 1.8 | 1.8 | 0.9 | 0 | 1.8 | 1.8 | 1.8 | 0.9 | 1.8 |
| P26 | 0.9 | 1.8 | 1.8 | 0.9 | 1.8 | 1.8 | 1.8 | 0.9 | 1.8 | 1.8 | 0 | 1.8 | 1.8 | 1.8 | 1.8 |
| P27 | 1.8 | 1.8 | 1.8 | 1.8 | 1.8 | 0.9 | 0.9 | 1.8 | 1.8 | 1.8 | 1.8 | 0 | 0.9 | 1.8 | 0.9 |
| P28 | 1.8 | 1.8 | 1.8 | 1.8 | 1.8 | 0.9 | 0.9 | 1.8 | 1.8 | 1.8 | 1.8 | 0.9 | 0 | 1.8 | 0.9 |
| P29 | 1.8 | 0.9 | 0.9 | 1.8 | 0.9 | 1.8 | 1.8 | 1.8 | 0.9 | 0.9 | 1.8 | 1.8 | 1.8 | 0 | 1.8 |
| P30 | 1.8 | 1.8 | 1.8 | 1.8 | 1.8 | 0.9 | 0.9 | 1.8 | 1.8 | 1.8 | 1.8 | 0.9 | 0.9 | 1.8 | 0 |

**Table E.3** Changeover times in stage S6 (h)

| | P01 | P02 | P03 | P04 | P05 | P06 | P07 | P08 | P09 | P10 | P11 | P12 | P13 | P14 | P15 |
|---|---|---|---|---|---|---|---|---|---|---|---|---|---|---|---|
| P01 | 0.0000 | 1.8000 | 1.2933 | 1.5183 | 1.5183 | 1.8000 | 1.8000 | 1.1250 | 0.9558 | 1.4625 | 1.8000 | 0.9558 | 1.8000 | 1.4625 | 1.1808 |
| P02 | 1.8000 | 0.0000 | 1.4625 | 1.4625 | 1.8000 | 1.8000 | 1.2933 | 1.1808 | 1.8000 | 1.8000 | 1.8000 | 0.8433 | 1.8000 | 1.2933 | 1.8000 |
| P03 | 1.2933 | 1.4625 | 0.0000 | 0.7875 | 1.4625 | 1.4625 | 1.4625 | 1.8000 | 1.2933 | 1.1250 | 1.4625 | 1.2933 | 1.4625 | 1.4625 | 1.8000 |
| P04 | 1.5183 | 1.4625 | 0.7875 | 0.0000 | 1.1808 | 0.9558 | 1.4625 | 1.2933 | 1.8000 | 1.1250 | 0.9558 | 1.8000 | 0.9558 | 1.4625 | 1.5183 |
| P05 | 1.5183 | 1.8000 | 1.4625 | 1.1808 | 0.0000 | 1.1250 | 1.1250 | 1.8000 | 1.4625 | 0.9558 | 1.4625 | 1.8000 | 1.4625 | 1.8000 | 1.0125 |
| P06 | 1.8000 | 1.8000 | 1.4625 | 0.9558 | 1.1250 | 0.0000 | 1.1250 | 1.2933 | 1.4625 | 1.4625 | 0.6183 | 1.8000 | 0.6183 | 1.8000 | 1.8000 |
| P07 | 1.8000 | 1.2933 | 1.4625 | 1.4625 | 1.1250 | 1.1250 | 0.0000 | 1.8000 | 1.4625 | 1.4625 | 1.4625 | 1.8000 | 1.4625 | 1.2933 | 1.4625 |
| P08 | 1.8000 | 1.1808 | 1.8000 | 1.2933 | 1.8000 | 1.2933 | 1.8000 | 0.0000 | 1.4625 | 1.4625 | 1.2933 | 0.8433 | 1.2933 | 1.4625 | 1.4625 |
| P09 | 0.9558 | 1.8000 | 1.2933 | 1.8000 | 1.4625 | 1.4625 | 1.4625 | 1.4625 | 0.0000 | 1.8000 | 1.5183 | 1.2933 | 1.8000 | 1.1808 | 1.8000 |
| P10 | 1.4625 | 1.8000 | 1.1250 | 1.1250 | 0.9558 | 1.4625 | 1.4625 | 1.4625 | 1.8000 | 0.0000 | 1.4625 | 1.8000 | 1.4625 | 1.4625 | 1.8000 |
| P11 | 1.8000 | 1.8000 | 1.4625 | 0.9558 | 1.4625 | 0.6183 | 1.4625 | 1.2933 | 1.5183 | 1.4625 | 0.0000 | 1.8000 | 0.3753 | 1.1308 | 1.8000 |
| P12 | 0.9558 | 0.8433 | 1.2933 | 1.8000 | 1.8000 | 1.8000 | 1.8000 | 0.8433 | 1.2933 | 1.4625 | 1.8000 | 0.0000 | 1.8000 | 1.8000 | 1.8000 |
| P13 | 1.8000 | 1.8000 | 1.4625 | 0.9558 | 1.4625 | 0.6183 | 1.4625 | 1.2933 | 1.8000 | 1.8000 | 0.3753 | 1.8000 | 0.0000 | 1.4625 | 1.8000 |
| P14 | 1.4625 | 1.2933 | 1.4625 | 1.4625 | 1.8000 | 1.8000 | 1.2933 | 1.4625 | 1.1808 | 1.4625 | 1.1808 | 1.8000 | 1.4625 | 0.0000 | 1.8000 |
| P15 | 1.1808 | 1.8000 | 1.8000 | 1.5183 | 1.0125 | 1.8000 | 1.4625 | 1.4625 | 1.8000 | 0.9558 | 0.9558 | 1.4625 | 1.8000 | 1.8000 | 0.0000 |
| P16 | 1.4625 | 1.8000 | 1.8000 | 1.2933 | 1.8000 | 0.6750 | 1.8000 | 0.9558 | 1.8000 | 1.4625 | 1.8000 | 1.4625 | 0.9558 | 1.8000 | 1.1250 |
| P17 | 1.1250 | 1.8000 | 1.8000 | 1.8000 | 0.9558 | 1.1808 | 1.4625 | 1.4625 | 1.1250 | 1.2933 | 1.8000 | 1.8000 | 1.8000 | 1.4625 | 1.2933 |
| P18 | 1.5183 | 1.8000 | 1.8000 | 1.0125 | 1.1808 | 0.9558 | 1.1250 | 1.2933 | 1.4625 | 1.8000 | 1.2933 | 1.8000 | 1.2933 | 1.8000 | 0.8433 |
| P19 | 1.8000 | 1.1250 | 1.1808 | 1.4625 | 1.2933 | 1.8000 | 1.8000 | 1.8000 | 1.4625 | 1.2933 | 1.8000 | 1.4625 | 1.8000 | 1.8000 | 1.2933 |
| P20 | 1.4625 | 1.8000 | 1.8000 | 1.8000 | 0.9558 | 1.4625 | 1.1250 | 1.4625 | 1.1250 | 1.2933 | 1.8000 | 1.8000 | 1.5183 | 1.4625 | 0.9558 |
| P21 | 1.2933 | 1.4625 | 0.6183 | 1.1250 | 1.4625 | 1.4625 | 1.4625 | 1.8000 | 0.9558 | 1.4625 | 1.4625 | 1.2933 | 1.4625 | 1.8000 | 1.8000 |
| P22 | 1.8000 | 1.4625 | 1.8000 | 1.2933 | 1.4625 | 1.0125 | 1.8000 | 1.2933 | 1.8000 | 1.8000 | 0.9558 | 1.4625 | 0.9558 | 1.4625 | 1.8000 |
| P23 | 1.4625 | 1.8000 | 1.1250 | 0.6183 | 1.4625 | 0.9558 | 1.4625 | 0.9558 | 1.5183 | 0.7875 | 0.6750 | 1.4625 | 0.9558 | 1.1808 | 1.4625 |
| P24 | 1.4625 | 1.2933 | 1.4625 | 1.4625 | 1.4625 | 1.4625 | 0.4500 | 1.4625 | 1.8000 | 1.1250 | 1.4625 | 1.4625 | 1.4625 | 1.2933 | 1.1250 |

(continued)

**Table E.3** (continued)

| | P01 | P02 | P03 | P04 | P05 | P06 | P07 | P08 | P09 | P10 | P11 | P12 | P13 | P14 | P15 |
|---|---|---|---|---|---|---|---|---|---|---|---|---|---|---|---|
| P25 | 1.4625 | 1.4625 | 1.1250 | 0.6183 | 1.8000 | 1.0125 | 1.8000 | 0.9558 | 1.4625 | 1.4625 | 1.2933 | 1.8000 | 1.2933 | 1.1250 | 1.8000 |
| P26 | 1.2933 | 1.8000 | 0.9558 | 1.4625 | 1.8000 | 1.8000 | 1.5183 | 1.8000 | 1.2933 | 1.4625 | 1.4625 | 1.2933 | 1.4625 | 1.1250 | 1.4625 |
| P27 | 1.8000 | 1.0125 | 1.4625 | 1.4625 | 0.7875 | 1.1250 | 0.6183 | 1.5183 | 1.4625 | 1.4625 | 1.4625 | 1.5183 | 1.4625 | 1.2933 | 1.8000 |
| P28 | 1.4625 | 1.1250 | 1.4625 | 0.9558 | 1.8000 | 1.2933 | 1.8000 | 0.6183 | 1.4625 | 1.8000 | 1.2933 | 1.4625 | 1.2933 | 1.4625 | 1.8000 |
| P29 | 1.4625 | 1.8000 | 1.4625 | 1.4625 | 1.2933 | 1.5183 | 1.8000 | 1.4625 | 1.8000 | 0.6183 | 1.8000 | 1.4625 | 1.8000 | 1.4625 | 0.6183 |
| P30 | 0.6750 | 1.4625 | 1.2933 | 1.5183 | 1.1808 | 1.8000 | 1.8000 | 1.4625 | 1.2933 | 1.4625 | 1.8000 | 0.6183 | 1.8000 | 1.8000 | 1.1808 |

| | P16 | P17 | P18 | P19 | P20 | P21 | P22 | P23 | P24 | P25 | P26 | P27 | P28 | P29 | P30 |
|---|---|---|---|---|---|---|---|---|---|---|---|---|---|---|---|
| P01 | 1.4625 | 1.1250 | 1.5183 | 1.8000 | 1.4625 | 1.2933 | 1.8000 | 1.4625 | 1.4625 | 1.4625 | 1.2933 | 1.8000 | 1.4625 | 1.4625 | 0.6750 |
| P02 | 1.8000 | 1.8000 | 1.8000 | 1.1250 | 1.8000 | 1.4625 | 1.4625 | 1.8000 | 1.2933 | 1.4625 | 1.8000 | 1.0125 | 1.1250 | 1.8000 | 1.4625 |
| P03 | 1.8000 | 1.8000 | 1.8000 | 1.1808 | 1.8000 | 0.6183 | 1.8000 | 1.1250 | 1.4625 | 1.1250 | 0.9558 | 1.4625 | 1.4625 | 1.4625 | 1.2933 |
| P04 | 1.2933 | 1.8000 | 1.0125 | 1.4625 | 1.8000 | 1.1250 | 1.2933 | 0.6183 | 1.4625 | 0.6183 | 1.4625 | 1.4625 | 0.9558 | 1.4625 | 1.5183 |
| P05 | 1.8000 | 0.9558 | 1.1808 | 1.2933 | 0.9558 | 1.4625 | 1.4625 | 1.4625 | 1.4625 | 1.8000 | 1.8000 | 0.7875 | 1.8000 | 1.2933 | 1.1803 |
| P06 | 0.6750 | 1.1808 | 0.9558 | 1.8000 | 1.4625 | 1.4625 | 1.0125 | 0.9558 | 1.4625 | 1.0125 | 1.8000 | 1.1250 | 1.2933 | 1.5183 | 1.8000 |
| P07 | 1.8000 | 1.4625 | 1.1250 | 1.8000 | 1.1250 | 1.4625 | 1.8000 | 1.4625 | 0.4500 | 1.8000 | 1.5183 | 0.6183 | 1.8000 | 1.8000 | 1.8000 |
| P08 | 0.9558 | 1.4625 | 1.2933 | 1.8000 | 1.4625 | 1.8000 | 1.2933 | 0.9558 | 1.4625 | 0.9558 | 1.8000 | 1.5183 | 0.6183 | 1.4625 | 1.4625 |
| P09 | 1.8000 | 1.1250 | 1.4625 | 1.8000 | 1.1250 | 0.9558 | 1.8000 | 1.5183 | 1.8000 | 1.4625 | 1.2933 | 1.4625 | 1.4625 | 1.8000 | 1.2933 |
| P10 | 1.4625 | 1.2933 | 1.8000 | 1.2933 | 1.2933 | 1.4625 | 1.8000 | 0.7875 | 1.1250 | 1.2933 | 1.4625 | 1.4625 | 1.8000 | 0.6183 | 1.4625 |
| P11 | 0.9558 | 1.8000 | 1.2933 | 1.8000 | 1.8000 | 1.4625 | 0.9558 | 0.6750 | 1.4625 | 1.8000 | 1.4625 | 1.4625 | 1.2933 | 1.8000 | 1.8000 |
| P12 | 1.4625 | 1.8000 | 1.8000 | 1.4625 | 1.8000 | 1.2933 | 1.4625 | 1.4625 | 1.4625 | 1.0125 | 1.2933 | 1.5183 | 1.4625 | 1.4625 | 1.8000 |
| P13 | 0.9558 | 1.8000 | 1.2933 | 1.8000 | 1.5183 | 1.4625 | 0.9558 | 0.9558 | 1.4625 | 1.8000 | 1.4625 | 0.6183 | 1.2933 | 1.8000 | 1.8000 |
| P14 | 1.8000 | 1.4625 | 1.8000 | 1.8000 | 1.4625 | 1.8000 | 1.4625 | 1.1808 | 1.2933 | 1.2933 | 1.1250 | 1.2933 | 1.4625 | 1.4625 | 1.4625 |
| P15 | 1.1250 | 1.2933 | 0.8433 | 1.2933 | 0.9558 | 1.8000 | 1.8000 | 1.4625 | 1.1250 | 1.8000 | 1.4625 | 1.8000 | 1.8000 | 0.6183 | 1.1803 |
| P16 | 0.0000 | 1.5183 | 0.9558 | 1.8000 | 1.8000 | 1.8000 | 1.0125 | 0.9558 | 1.4625 | 1.0125 | 1.4625 | 1.8000 | 1.2933 | 0.8433 | 1.4625 |
| P17 | 1.5183 | 0.0000 | 1.4625 | 1.2933 | 0.6183 | 1.8000 | 1.5183 | 1.8000 | 1.8000 | 1.1808 | 1.8000 | 1.4625 | 1.4625 | 1.0125 | 1.8000 |

(continued)

**Table E.3** (continued)

|      | P16    | P17    | P18    | P19    | P20    | P21    | P22    | P23    | P24    | P25    | P26    | P27    | P28    | P29    | P30    |
|------|--------|--------|--------|--------|--------|--------|--------|--------|--------|--------|--------|--------|--------|--------|--------|
| P18  | 0.9558 | 1.4625 | 0.0000 | 1.8000 | 1.1250 | 1.8000 | 1.2933 | 1.2933 | 1.4625 | 1.2933 | 1.4625 | 1.4625 | 1.2933 | 1.4625 | 1.5183 |
| P19  | 1.8000 | 1.2933 | 1.8000 | 0.0000 | 1.2933 | 1.1250 | 1.4625 | 1.8000 | 1.8000 | 1.4625 | 1.8000 | 1.8000 | 1.4625 | 1.2933 | 1.4625 |
| P20  | 1.8000 | 0.6133 | 1.1250 | 1.2933 | 0.0000 | 1.8000 | 1.8000 | 1.8000 | 1.4625 | 1.4625 | 1.8000 | 1.4625 | 1.4625 | 1.2933 | 1.8000 |
| P21  | 1.8000 | 1.8000 | 1.8000 | 1.1250 | 1.8000 | 0.0000 | 1.8000 | 1.4625 | 1.4625 | 1.4625 | 1.2933 | 1.4625 | 1.1808 | 1.8000 | 1.2933 |
| P22  | 1.0125 | 1.5183 | 1.2933 | 1.4625 | 1.8000 | 1.8000 | 0.0000 | 1.2933 | 1.8000 | 1.0125 | 1.4625 | 1.4625 | 1.2933 | 1.5183 | 1.1250 |
| P23  | 0.9558 | 1.8000 | 1.2933 | 1.8000 | 1.8000 | 1.4625 | 1.2933 | 0.0000 | 1.1250 | 0.9558 | 1.4625 | 1.4625 | 1.2933 | 1.1250 | 1.4625 |
| P24  | 1.4625 | 1.8000 | 1.4625 | 1.8000 | 1.4625 | 1.4625 | 1.8000 | 1.1250 | 0.0000 | 1.8000 | 1.5183 | 0.9558 | 1.8000 | 1.4625 | 1.4625 |
| P25  | 1.0125 | 1.1808 | 1.2933 | 1.4625 | 1.4625 | 1.4625 | 1.0125 | 0.9558 | 1.8000 | 0.0000 | 1.4625 | 1.8000 | 0.6183 | 1.1808 | 1.8000 |
| P26  | 1.4625 | 1.8000 | 1.4625 | 1.8000 | 1.8000 | 1.2933 | 1.4625 | 1.4625 | 1.5183 | 0.0000 | 0.0000 | 1.8000 | 1.8000 | 1.1250 | 1.2933 |
| P27  | 1.8000 | 1.4625 | 1.4625 | 1.8000 | 1.4625 | 1.4625 | 1.4625 | 1.4625 | 0.9558 | 1.8000 | 1.8000 | 0.0000 | 1.8000 | 1.8000 | 1.4625 |
| P28  | 1.2933 | 1.4625 | 1.2933 | 1.4625 | 1.4625 | 1.1808 | 1.2933 | 1.2933 | 1.8000 | 0.6183 | 1.8000 | 1.8000 | 0.0000 | 1.8000 | 1.8000 |
| P29  | 0.8433 | 1.0125 | 1.4625 | 1.2933 | 1.2933 | 1.8000 | 1.5183 | 1.1250 | 1.4625 | 1.1808 | 1.1250 | 1.8000 | 1.8000 | 0.0000 | 1.4625 |
| P30  | 1.4625 | 1.8000 | 1.5183 | 1.4625 | 1.8000 | 1.2933 | 1.1250 | 1.4625 | 1.4625 | 1.8000 | 1.2933 | 1.4625 | 1.8000 | 1.4625 | 0.0000 |

**Table E.4** Processing times (h)

| Product | M01 | M02 | M03 | M04 | M05 | M06 | M07 | M08 | M09 | M10 | M11 | M12 | M13 | M14 | M15 | M16 | M17 |
|---|---|---|---|---|---|---|---|---|---|---|---|---|---|---|---|---|---|
| P01 | 0.9000 | 0.9000 | 1.3050 | 1.3050 | 2.0979 | 1.6335 | 2.0205 | 2.0205 | – | – | 0.5778 | 0.2250 | 0.3780 | 0.3780 | 1.9818 | 1.9818 | 0.5661 |
| P02 | 0.9000 | 0.9000 | 1.3050 | 1.3050 | – | – | – | – | – | – | 0.6894 | 0.2250 | 0.3780 | 0.3780 | 2.3634 | 2.3634 | 0.6750 |
| P03 | 0.9000 | 0.9000 | 1.3050 | 1.3050 | – | – | – | – | 0.3339 | 0.3339 | – | 0.6750 | 1.1340 | 1.1340 | 0.3276 | 0.3276 | 0.0936 |
| P04 | 0.9000 | 0.9000 | 1.3050 | 1.3050 | – | – | – | – | – | – | 0.5832 | 0.4500 | 0.7560 | 0.7560 | 1.9998 | 1.9998 | 0.5715 |
| P05 | 0.9000 | 0.9000 | 1.3050 | 1.3050 | – | – | – | – | 0.2223 | 0.2223 | – | 0.2250 | 0.3780 | 0.3780 | 0.2178 | 0.2178 | 0.0621 |
| P06 | 0.9000 | 0.9000 | 1.3050 | 1.3050 | – | – | – | – | 2.0403 | 2.0403 | 0.5832 | 0.2250 | 0.3780 | 0.3780 | 1.9998 | 1.9998 | 0.5715 |
| P07 | 0.9000 | 0.9000 | 1.3050 | 1.3050 | – | – | – | – | – | – | 0.1062 | 0.2250 | 0.3780 | 0.3780 | 0.3636 | 0.3636 | 0.1035 |
| P08 | 0.9000 | 0.9000 | 1.3050 | 1.3050 | 2.0979 | 1.6335 | 2.0205 | 2.0205 | – | 0.3708 | 0.1062 | 0.6750 | 1.1340 | 1.1340 | 0.3636 | 0.3636 | 0.1035 |
| P09 | 0.9000 | 0.9000 | 1.3050 | 1.3050 | – | – | – | – | – | – | 0.1008 | 0.6750 | 1.1340 | 1.1340 | 0.3456 | 0.3456 | 0.0990 |
| P10 | 0.9000 | 0.9000 | 1.3050 | 1.3050 | – | – | – | – | – | – | 0.2646 | 0.4500 | 0.7560 | 0.7560 | 0.9090 | 0.9090 | 0.2601 |
| P11 | 0.9000 | 0.9000 | 1.3050 | 1.3050 | – | – | – | – | – | – | 0.7947 | 0.4500 | 0.7560 | 0.7560 | 2.7270 | 2.7270 | 0.7794 |
| P12 | 0.9000 | 0.9000 | 1.3050 | 1.3050 | 2.0979 | 1.6335 | 2.0205 | 2.0205 | – | 1.8849 | – | 0.4500 | 0.7560 | 0.7560 | 1.8180 | 1.8180 | 0.5193 |
| P13 | 0.9000 | 0.9000 | 1.3050 | 1.3050 | 2.0979 | 1.6335 | 2.0205 | 2.0205 | 0.3528 | 0.7416 | – | 0.2250 | 0.3780 | 0.3780 | 0.7272 | 0.7272 | 0.2079 |
| P14 | 0.9000 | 0.9000 | 1.3050 | 1.3050 | – | – | – | – | 0.3528 | 0.3528 | 0.1008 | 0.4500 | 0.7560 | 0.7560 | 0.3456 | 0.3456 | 0.0990 |
| P15 | 0.9000 | 0.9000 | 1.3050 | 1.3050 | – | – | – | – | 2.0403 | 2.0403 | 0.5832 | 0.4500 | 0.7560 | 0.7560 | 1.9998 | 1.9998 | 0.5715 |
| P16 | 0.9000 | 0.9000 | 1.3050 | 1.3050 | 2.0979 | 1.6335 | 2.0205 | 2.0205 | – | – | 0.2124 | 0.4500 | 0.7560 | 0.7560 | 0.7272 | 0.7272 | 0.2079 |
| P17 | 0.9000 | 0.9000 | 1.3050 | 1.3050 | 2.0979 | 1.6335 | 2.0205 | 2.0205 | 1.1133 | 1.1133 | – | 0.4500 | 0.7560 | 0.7560 | 1.0908 | 1.0908 | 0.3114 |
| P18 | 0.9000 | 0.9000 | 1.3050 | 1.3050 | – | – | – | – | – | – | 0.6363 | 0.4500 | 0.7560 | 0.7560 | 2.1816 | 2.1816 | 0.6237 |
| P19 | 0.9000 | 0.9000 | 1.3050 | 1.3050 | – | – | – | – | – | – | 0.5616 | 0.4500 | 0.7560 | 0.7560 | 1.9269 | 1.9269 | 0.5508 |
| P20 | 0.9000 | 0.9000 | 1.3050 | 1.3050 | 2.0979 | 1.6335 | 2.0205 | 2.0205 | – | 0.3339 | 0.0954 | 0.6750 | 1.1340 | 1.1340 | 0.3276 | 0.3276 | 0.0936 |
| P21 | 0.9000 | 0.9000 | 1.3050 | 1.3050 | – | – | 2.0205 | 2.0205 | – | 0.2781 | – | 0.6750 | 1.1340 | 1.1340 | 0.2727 | 0.2727 | 0.0783 |
| P22 | 0.9000 | 0.9000 | 1.3050 | 1.3050 | – | – | – | – | – | – | 0.5409 | 0.6750 | 1.1340 | 1.1340 | 1.8549 | 1.8549 | 0.5301 |
| P23 | 0.9000 | 0.9000 | 1.3050 | 1.3050 | 2.0979 | 1.6335 | 2.0205 | 2.0205 | – | 2.0403 | 0.5832 | 0.6750 | 1.1340 | 1.1340 | 1.9998 | 1.9998 | 0.5715 |
| P24 | 0.9000 | 0.9000 | 1.3050 | 1.3050 | – | | 2.0205 | 2.0205 | – | 2.2257 | 0.6363 | 0.2250 | 0.3780 | 0.3780 | 2.1816 | 2.1816 | 0.6237 |
| P25 | 0.9000 | 0.9000 | 1.3050 | 1.3050 | – | – | – | – | – | 1.1133 | – | 0.6750 | 1.1340 | 1.1340 | 1.0908 | 1.0908 | 0.3114 |
| P26 | 0.9000 | 0.9000 | 1.3050 | 1.3050 | 2.0979 | 1.6335 | 2.0205 | 2.0205 | – | 1.9476 | 0.5562 | 0.4500 | 0.7560 | 0.7560 | 1.9089 | 1.9089 | 0.5454 |

(continued)

**Table E.4** (continued)

| Product | M01 | M02 | M03 | M04 | M05 | M06 | M07 | M08 | M09 | M10 | M11 | M12 | M13 | M14 | M15 | M16 | M17 |
|---|---|---|---|---|---|---|---|---|---|---|---|---|---|---|---|---|---|
| P27 | 0.9000 | 0.9000 | 1.3050 | 1.3050 | – | – | 2.0205 | 2.0205 | – | 2.0403 | 0.5832 | 0.2250 | 0.3780 | 0.3780 | 1.9998 | 1.9998 | 0.5715 |
| P28 | 0.9000 | 0.9000 | 1.3050 | 1.3050 | – | – | – | – | – | – | 0.1062 | 0.2250 | 0.3780 | 0.3780 | 0.3636 | 0.3636 | 0.1035 |
| P29 | 0.9000 | 0.9000 | 1.3050 | 1.3050 | – | – | – | – | 1.6695 | 1.6695 | – | 0.2250 | 0.3780 | 0.3780 | 1.6362 | 1.6362 | 0.4671 |
| P30 | 0.9000 | 0.9000 | 1.3050 | 1.3050 | – | – | – | – | – | 3.3390 | – | 0.2250 | 0.3780 | 0.3780 | 3.2724 | 3.2724 | 0.9351 |

# Bibliography

Neumann K, Schwindt C, Trautmann N (2002) Advanced production scheduling for batch plants in process industries. OR Spectrum 24:251–279

Papageorgiou LG (2009) Supply chain optimisation for the process industries: advances and opportunities. Comput Chem Eng 33(12):1931–1938

Sung C, Maravelias CT (2007) An attainable region approach for production planning of multi-product processes. AIChE J 53(5):1298–1315

© Springer Nature Switzerland AG 2019    291
G. M. Kopanos and L. Puigjaner, *Solving Large-Scale Production
Scheduling and Planning in the Process Industries*,
https://doi.org/10.1007/978-3-030-01183-3

Printed in the United States
By Bookmasters